Isaac Asimov, world maestro of science fiction, was born in Russia near Smolensk in 1920 and brought to the United States by his parents three years later. He grew up in Brooklyn where he went to grammar school and at the age of eight he gained his citizen papers. A remarkable memory helped him to finish high school before he was sixteen. He then went on to Columbia University, resolved to become a chemist rather than follow the medical career his father had in mind for him. He graduated in Chemistry and after a short spell in the Army he gained his doctorate in 1949 and qualified as an instructor in biochemistry at Boston University School of Medicine where he became associate professor in 1955, doing research in nucleic acid. Increasingly, however, the pressures of chemical researches conflicted with his aspirations in the literary field, and in 1958, he retired to full-time authorship while retaining his connection with the university.

Asimov's fantastic career as a science fiction writer began in 1939 with the appearance of a short story *Marooned Off Vesta* in *Amazing Stories*. Thereafter he became a regular contributor to the leading SF magazines of the day including: *Astounding, Astonishing Stories, Super Science Stories* and *Galaxy*. Many of his most famous novels have also appeared in serialization form. With over one hundred and fifty books to his credit and several hundred articles, Asimov's output is prolific by any standards. He has won the Hugo Award three times and the Nebula Award once. Apart from his many world-famous science fiction works, he has also written detective mystery stories, a four-volume *History of North America*, a two-volume *Guide to the Bible*, a biographical dictionary, encyclopaedias, textbooks, as well as a long and impressive list of books on many aspects of science.

# Also by Isaac Asimov

*Foundation*
*Foundation and Empire*
*Second Foundation*

*Earth is Room Enough*
*The Stars Like Dust*
*The Martian Way*
*The Currents of Space*
*The End of Eternity*
*The Naked Sun*
*The Caves of Steel*
*Asimov's Mysteries*
*The Gods Themselves*
*Nightfall One*
*Nightfall Two*

*I, Robot*
*The Rest of the Robots*

*The Early Asimov: Volume I*
*The Early Asimov: Volume II*
*The Early Asimov: Volume III*

*Nebula Award Stories 8* (editor)
*The Stars in their Courses* (non-fiction)
*Tales of the Black Widowers* (detection)

Isaac Asimov

# The Left Hand of the Electron

Panther

Granada Publishing Limited
Published in 1976 by Panther Books Ltd
Frogmore, St Albans, Herts AL2 2NF

First published in Great Britain by White Lion
Publishers Ltd 1975
Copyright © Isaac Asimov 1972
All other essays copyright © Mercury Press, Inc
1970, 1971, 1972
Made and printed in Great Britain by
Cox & Wyman Ltd, London, Reading and Fakenham

This book is sold subject to the condition that it shall not,
by way of trade or otherwise, be lent, re-sold, hired out
or otherwise circulated without the publisher's prior
consent in any form of binding or cover other than that
in which it is published and without a similar condition
including this condition being imposed on the
subsequent purchaser.
This book is published at a net price and is supplied
subject to the Publishers Association Standard
Conditions of Sale registered under the Restrictive
Trade Practices Act, 1956.

**Dedicated to:**
*Rae, as a book of her own*

# CONTENTS

INTRODUCTION 9

## A – *The Problem of Left and Right*

1 – ODDS AND EVENS 14
2 – THE LEFT HAND OF THE ELECTRON 26
3 – SEEING DOUBLE 39
4 – THE 3-D MOLECULE 52
5 – THE ASYMMETRY OF LIFE 65

## B – *The Problem of Oceans*

6 – THE THALASSOGENS 78
7 – HOT WATER 91
8 – COLD WATER 104

## C – *The Problem of Numbers and Lines*

9 – PRIME QUALITY 118
10 – EUCLID'S FIFTH 130
11 – THE PLANE TRUTH 143

## D – *The Problem of the Platypus*

12 – HOLES IN THE HEAD 158

## E – *The Problem of History*

    13 – THE EUREKA PHENOMENON      172

    14 – POMPEY AND CIRCUMSTANCE      185

    15 – BILL AND I      199

## F – *The Problem of Population*

    16 – STOP!      212

    17 – ... BUT HOW?      225

# INTRODUCTION

I feel slightly embarrassed to admit that one of the chapters in this book (the fifth one, to be specific) is actually my 160th monthly essay for *The Magazine of Fantasy and Science Fiction*. For over thirteen years, I have been writing an essay each month, without fail, for that noble periodical and I live in constant terror that the day may be coming when I'll hear in not-so-dulcet tones that horrible phrase 'Enough already!!!' with at least three exclamation points.

But from whom? Certainly not from myself, since in all those months, though I have written all kinds of other items, from history to jokebooks and from science fiction to biblical commentary, the one piece of writing to which I have consistently turned most eagerly is my monthly *F&SF* essay. Why? Because— Well, I'll explain in a little while.

Well then, will the demand for a halt come from the gentle editor of *F&SF*? Never, I hope. At least, he assures me it will be never.

But there are other editors involved, too. This is the ninth volume of collected *F&SF* essays, each one of which has been put out with absolute and smiling loyalty by Doubleday & Co. just as fast as I bring them in. And *they* assure me they won't stop, either.

Will it be the readers then? They may get tired, I suppose, and stop shelling out hard-earned money to read my infernal cackle of enthusiasm over this point and that for chapter after chapter after chapter and volume after volume after volume. I suppose that's possible. I hope none of you will ever grow tired of me but this is something I don't have control over. You may!

And if you do, Doubleday & Co. will have to come to me

and explain to me the harsh facts of economic life. (I can see it now, in my mind's eye— A meeting of editors; the pulling of the short straw to see who gets to tell me; the embarrassed clearing of the throat; the sentence that begins, 'Gee, Isaac, it's this way—')

And then a quiet word from the editor of *F&SF* to the effect that in order to build circulation it will be necessary to—

And then what?

Well, I'll tell you. If you all fail me (as let's hope you never do) I will keep on writing the essays anyway; one a month at least. I couldn't help it, because by now they have become my standard way of decreasing internal pressures and relieving the possibility of explosion.

If I brood over the world's problems, there is the alternative of staying awake night after night (for sleep and I have been lifelong enemies, anyway) or of sitting down to write the essays 'Stop!' and '—But How?'

If a Shakespearian scholar lifts two dainty fingers to his patrician nose when a lowly non-Shakespearian dares invade the sacred precincts and write a two-volume work entitled *Asimov's Guide to Shakespeare*, I can, of course, in the time-honored fashion of writers throughout history, throw a screaming tantrum and go shopping for horsewhips. Or I can do as I actually did, and write 'Bill and I', getting rid of my spleen and, at the same time, making (I honestly believe) a valid point.

If I come across an interesting and far-out coincidence in writing one of my other books – say, one on Roman history – I don't have to let it lie there. I can get to work and pound the thing firmly into the ground in 'Pompey and Circumstance', managing to get off, at the same time, a title containing one of the noblest puns I have ever invented in a lifetime of inventing noble puns.

(I once told the story of the braggart who claimed to be the best surfer in the world. Finally, on a beautiful beach with magnificent waves, a surfboard was thrust into his hands and he was told to go out and demonstrate. He marched down the beach to just above the high-water line, stuck the surfboard upright into the sand and stood there motionless. 'Go ahead,' everyone yelled, 'get into the water.' But he shouted back, 'I don't have to. They also surf who only stand and wait.'

I barely got away with my life, that time.)

Or else I will come across a short news article in a science magazine to the effect that a certain line of research seems to draw a connection between the non-conservation of parity and the optical activity of naturally occurring compounds. The whole article took up about five hundred words.

At once I am on fire to explain this to *my* readers in *my* way. Of course, that means that first I have to explain about parity; how it is conserved and how it is not conserved; which takes up two articles of four thousand words each. Then I have to explain all that jazz about optical activity, which takes up two more articles. *Then* I combine the two in a fifth article and here is the whole bunch (pant, pant) as the first five chapters of this book.

Or someone in the audience stops me after one of my talks and says, 'Why don't you ever write an article on non-Euclidean geometry?' and I say, affably, 'Good idea!' and then it plagues me and plagues me and I write not one but two such articles, 'Euclid's Fifth' and 'The Plane Truth'. —Then I can breathe again.

But the prize example happened one time when I rebelled and said to myself, 'Goodness gracious' (for I use strong language when moved and that is my favorite) 'I'm not going to write today. For once, I'm getting into bed and just do nothing all day but read.'

So I did. I got a math book that I had always wanted to read in a leisurely way when I had nothing else to do. I pulled down all the window shades (I hate daylight), turned on the headlamp, got into bed, and began reading. The first chapter was on prime numbers Before I finished, I began thinking – and thinking—

And I got out of bed in a white-hot fever of impatience

and wrote 'Prime Quality' and never did get to spending the day in bed.

So how will any of this stop just because the readers don't want any more and because the editors will then be forced, by the law of the marketplace, to go along with them?

I will just have to continue writing the essays for myself alone, and pile them up one on top of the other, and every once in a while leaf through them and read them and reread them and enjoy them all by myself.

'Enough already?'

For you, maybe. But for me: 'No! No!! Never!!!'

# A – The Problem of Left and Right

# 1 – ODDS AND EVENS

I have just gone through a rather unsettling experience. Ordinarily it is not very difficult to think up a topic for these chapters. Some interesting point will occur to me, which will quickly lead my mind to a particular line of development, beginning in one place and ending in another. Then, I get started.

Today, however, having determined to deal with asymmetry (in more than one chapter, very likely) and to end with life and antilife, I found that two possible starting points occurred to me. Ordinarily, when this happens, one starting point seems so much superior to me that I choose it over the other with a minimum of hesitation.

This time, however, the question was whether to start with even numbers or with double refraction, and the arguments raging within my head for each case were so equally balanced that I couldn't make up my mind. For two hours I sat at my desk, pondering first one and then the other and growing steadily unhappier.

Indeed, I became uncomfortably aware of the resemblance of my case to that of 'Buridan's ass'.

The reference, here, is to a fourteenth-century French philosopher, Jean Buridan, who was supposed to have stated the following: 'If a hungry ass were placed exactly between two haystacks in every respect equal, it would starve to death, because there would be no motive why it should go to one rather than to the other.'

Actually, of course, there's a fallacy here, since the statement does not recognize the existence of the random factor. The ass, no logician, is bound to turn his head randomly so that one haystack comes into better view, shuffle his feet randomly so that one haystack comes to be closer; and he would end at the haystack better seen or more closely approached.

*Which* haystack that would be, one could not tell in advance. If one had a thousand asses placed exactly between a thousand sets of haystack pairs, one could confidently expect that about half would turn to the right and half to the left. The individual ass, however, would remain unpredictable.

In the same way, it is impossible to predict whether an honest coin, honestly thrown, will come down heads or tails in any one particular case, but we can confidently predict that a very large number of coins tossed simultaneously (or one coin tossed a very large number of times) will show heads just about half the time and tails the other half.

And so it happens that although the chance of the fall of heads or tails is exactly even, just fifty-fifty, you can nevertheless call upon the aid of randomness to help you make a decision by tossing one coin once.

At this point, I snapped out of my reverie and did what a lesser mind would have done two hours before. I tossed a coin.

Shall we start with even numbers, Gentle Readers?

I suspect that some prehistoric philosopher must have decided that there were two kinds of numbers: peaceful ones and warlike ones. The peaceful numbers were those of the type 2, 4, 6, 8, while the numbers in between were warlike.

If there were 8 stone axes and two individuals possessing equal claim, it would be easy to hand 4 to each and make peace. If there were 7, however, you would have to give 3 to each and then either toss away the 1 remaining (a clear loss of a valuable object) or let the two disputants fight over it.

The fact that the original property that marked out the significance of what we now call even and odd numbers was something like this is indicated by the very names we give them.

The word 'even' means fundamentally, 'flat, smooth, without unusual depressions or elevations'. We use the word in this sense when we say that a person says something 'in an even tone of voice'. An even number of identical coins, for instance, can be divided into two piles of exactly the same height. The two piles are even in height and hence the number is called even. The even number is the one with the property of 'equal shares'.

'Odd', on the other hand, is from an old Norse word meaning 'point' or 'tip'. If an odd number of coins is divided into two piles as nearly equal as possible, one pile is higher by one coin and therefore rears a point or tip into the air, as compared with the other. The odd number possesses the property of 'unequal shares', and it is no accident that the expression 'odds' in betting implies the wagering of unequal amounts of money by the two participants.

Since even numbers possess the property of equal shares, they were said to have 'parity', from a Latin word meaning 'equal'. Originally, this word applied (as logic demanded) to even numbers only, but mathematicians found it convenient to say that if two numbers were both even or both odd, they were, in each case 'of the same parity'. An even number and an odd number, grouped together, were 'of different parity'.

To see the convenience of this convention, consider the following:

If two even numbers are added, the sum is invariably even. (This can be expressed mathematically by saying that two even numbers can be expressed as $2m$ and $2n$ where $m$ and $n$ are whole numbers and that the sum, $2m+2n$, is still clearly divisible by two. However, we are friends, you and I, and I'm sure we can dispense with mathematical reasoning and that I will find you willing to accept my word of honor as a gentleman in such matters. Besides, you are welcome to search for two even numbers whose sum isn't even.)

If two odd numbers are added, the sum is also invariably even.

If an odd number and an even number are added, however, the sum is invariably odd.

We can express this more succinctly in symbols, with $E$ standing for even and $O$ standing for odd:

$$E + E = E$$
$$O + O = E$$
$$E + O = O$$
$$O + E = O$$

Or, if we are dealing with pairs of numbers only, the concept of parity enables us to say it in two statements, rather than four:

(1) Same parities add to even.
(2) Different parities add to odd.

A very similar state of affairs exists with reference to multiplication, if we divide numbers into two classes: positive numbers ($+$) and negative numbers ($-$). The product of two positive numbers is invariably positive. The product of two negative numbers is invariably positive. The product of a positive and a negative number is invariably negative. Using symbols:

$$+ \times + = +$$
$$- \times - = +$$
$$+ \times - = -$$
$$- \times + = -$$

Or, if we consider all positive numbers as having one kind of parity and all negative numbers as another, we can say, in connection with the multiplication of two numbers:

(1) Same parities multiply to positive.
(2) Different parities multiply to negative.

The concept of parity – that is, the assignment of all objects of a particular class to one of two subclasses and then finding two opposing results when objects of the same or of different subclasses are manipulated – can be applied to physical phenomena.

For instance, all electrically charged particles can be divided into two classes: positively charged and negatively charged. Again, all magnets possess two points of concentrated magnetism of opposite properties: a north pole and a south pole. Let's symbolize these as +, −, N and S.

It turns out that:

$$+ \text{ and } + \text{ or } N \text{ and } N = \text{repulsion}$$
$$- \text{ and } - \text{ or } S \text{ and } S = \text{repulsion}$$
$$+ \text{ and } - \text{ or } N \text{ and } S = \text{attraction}$$
$$- \text{ and } + \text{ or } S \text{ and } N = \text{attraction}$$

Again, we can make two statements:

(1) Like electric charges, or magnetic poles, repel each other.

(2) Opposite electric charges, or magnetic poles, attract each other.

The similarity in form to the situation with respect to the summing of odd and even, or the multiplying of positive and negative, is obvious.

When, in any situation, same parities always yield one result and different parities always yield the opposite result, we say that 'parity is conserved'. If two even numbers sometimes added up to an odd number; or if a positive number multiplied by a negative one sometimes yielded a positive product; or if two positively charged objects sometimes attracted each other; or if a north magnetic pole sometimes repelled a south magnetic pole, we would say that, 'The law of conservation of parity is violated.'

Certainly in connection with numbers and with electromagnetic phenomena, no one has ever observed the law of conservation of parity to have been violated, and no one seriously expects to observe a case in the future.

What about other cases?

Well, electromagnetism involves a field. That is, any electrically charged particle, or any magnet, is surrounded by a volume of space within which its properties are made manifest on other objects of the same sort. The other objects are also surrounded by a volume of space within which their properties are made manifest on the original object. One

ODDS AND EVENS 19

speaks, therefore, of an 'electromagnetic interaction' involving pairs of objects carrying electric charge or magnetic poles.

Up through the first years of the twentieth century, the only other kind of interaction known was the gravitational.

At first blush, there seems no easy way of involving gravitation with parity. There is no way of dividing objects into two groups, one with one kind of gravitational property and the other with the opposite kind.

All objects of a given mass possess the same intensity of gravitational interaction of the same sort. Any two objects with mass attract each other. There seems no such thing as 'gravitational repulsion' (and, according to Einstein's General Theory of Relativity there *can't* be such a thing). It is as though, in gravity, we can say only that $E+E = E$ or $+ \times + = +$.

To be sure, there is a chance that in the field of subatomic physics there might be some objects with mass that possess the usual gravitational properties and other objects with mass that possess gravitational properties of the opposite kind ('antigravity'). In that case, the chances are that it would turn out that two antigravitational objects attract each other just as two gravitational objects do; but that an antigravitational and a gravitational object would repel each other. The situation with respect to the gravitational interaction would be the reverse of the electromagnetic one (like gravities would attract and unlike gravities would repel) but, allowing for that reversal, parity would still be conserved.

The trouble is, though, that the gravitational interaction is so much more feeble than the electromagnetic interaction that gravitational interactions of subatomic particles are as yet impossible to measure and a sub-tiny attraction can't be differentiated from a sub-tiny repulsion. —So the question of parity and the gravitational field remains in abeyance.

As the twentieth century wore on, it came to be recognized that the gravitational and electromagnetic interactions were not the only ones that existed. Subatomic particles involved something else. To be sure, electrons had negative charges and protons had positive charges and with respect to this, they behaved in accordance with the rules of electromagnetic interaction. There were other events in the subatomic world, however, that had nothing to do with electromagnetism. There was, for instance, some sort of interaction involving particles, whether with or without electric charge, that showed itself only in the super-close quarters to be found within the atomic nucleus.

Did this 'nuclear interaction' involve parity?

Every subatomic particle has a certain quantum-mechanical property which can be expressed in a form involving three quantities, $x$, $y$, and $z$. In some cases, it is possible to change the sign of all three quantities from positive to negative without changing the sign of the expression as a whole. Particles in which this is true are said to have 'even parity'. In other cases, changing the signs of the three quantities *does* change the sign of the entire expression and a particle of which this is true is said to have 'odd parity'.

Why even and odd? Well, an even-parity particle can break up into two even-parity particles or two odd-parity particles, but never into one even-parity plus one odd-parity. An odd-parity particle, on the other hand, can break up into an odd-parity particle plus an even-parity one, but never into two odd-parity particles or two even-parity particles. This is analogous to the way in which an even number can be the sum of two even numbers or of two odd numbers, but never the sum of an even number and an odd number, while an odd number can be the sum of an even number and an odd number, but can never be the sum of two even numbers or of two odd numbers.

But then a particle called the 'K-meson' was discovered. It was unstable and quickly broke down into 'pi-mesons'. Some K-mesons gave off two pi-mesons in breaking down and some gave off three pi-mesons and that was instantly disturbing. If a K-meson did one, it ought not to be able to do the other. Thus an even number can be the sum of two odd numbers ($10 = 3+7$) and an odd number can be the sum of three odd numbers (11 = 3+7+1), but no number can be the sum of two odd numbers in one case and three odd numbers in another. It would be like expecting a number to be both odd and even. It would, in short, represent a violation of the law of conservation of parity.

Physicists therefore reasoned there must be two kinds of K-meson; an even-parity variety ('theta-meson') that broke down to two pi-mesons, and an odd-parity variety ('tau-meson') that broke down to three pi-mesons.

This did not turn out to be an altogether satisfactory solution, since there seemed to be no possible distinction one could make between the theta-meson and the tau-meson *except* for the number of pi-mesons it broke down into. To invent a difference in parity for two particles identical in every other respect seemed poor practice.

By 1956, a few physicists had begun to wonder if it weren't possible that the law of conservation of parity might not be broken in some cases. If that were so, maybe it wouldn't be necessary to try to make a distinction between the theta-meson and the tau-meson.

The suggestion roused the interest of two young Chinese-American physicists at Columbia, Chen Ning Yang and Tsung Dao Lee, who took into consideration the following—

There is, as a matter of fact, not one nuclear interaction, but two. The one that holds protons and neutrons together within the nucleus is an extremely strong one, about 130 times as strong as the electromagnetic interaction, so it is called the 'strong nuclear interaction'.

There is a second, 'weak nuclear interaction' which is only about a hundred-trillionth the intensity of the strong nuclear interaction (but still some trillion-trillion times as intense as the unimaginably weak gravitational interaction).

This meant that there were four types of interaction in the universe (and there is some theoretical reason for arguing that a fifth of any sort cannot exist, but I would hate to commit myself to that): (1) strong nuclear, (2) electromagnetic, (3) weak nuclear, and (4) gravitational.

## THE PROBLEM OF LEFT AND RIGHT

We can forget about the gravitational interaction for reasons I mentioned earlier in the article. Of the other three, it had been well established by 1956 that parity was conserved in the strong nuclear interaction and in the electromagnetic interaction. Numerous cases of such conservation were known and the matter was considered settled.

No one, however, had ever systematically checked the weak nuclear interaction with respect to parity, and the breakdown of the K-meson involved a weak nuclear interaction. To be sure, all physicists assumed that parity was conserved in the weak nuclear interaction but that was only an assumption.

Yang and Lee published a paper pointing this out – and suggested experiments that might be performed to check whether the weak nuclear interactions conserved parity or not. Those experiments were quickly carried out and the Yang-Lee suspicion that parity would not be conserved was shown to be correct. There was very little delay in awarding them shares in the Nobel prize in physics in 1957, at which time Yang was thirty-four and Lee, thirty-one.

You might ask, of course, why parity should be conserved in some interactions and not in others – and might not be satisfied with the answer 'Because that's the way the universe is.'

Indeed, by concentrating too hard on those cases where parity *is* conserved, you might get the notion that it is impossible, inconceivable, unthinkable to deal with a case where it *isn't* conserved. If the conservation of parity is then shown not to hold in some cases, the notion arises that this is a tremendous revolution that throws the entire structure of science into a state of collapse.

None of that is so.

Parity is not so essential a part of everything that exists that it must be conserved in all places, at all times, and under all conditions. Why shouldn't there be conditions where it isn't conserved or, as in the case of gravitational interaction, where it might not even apply?

It is also important to understand that the discovery of the fact that parity was not conserved in weak nuclear interactions did not 'overthrow' the law of conservation of parity, even though that was certainly the way in which it was presented in the newspapers and even by scientists themselves. The law of conservation of parity, in those cases in which its validity had been tested by experiment, *remained* and is still as much in force as ever.

It was only in connection with the weak nuclear interactions, where the validity of the law of conservation of parity *had never been tested* prior to 1956 and where it had merely been rather carelessly *assumed* that it applied, that there came the change. The final experiment merely showed that physicists had made an assumption they had no real right to make and the law of conservation of parity was 'overthrown' only where it had never been shown to exist in the first place.

It might help if we look at some familiar, everyday case where a law of conservation of parity holds, and then on another where it is merely assumed to hold by analogy, but doesn't really. We can then see what happened in physics, and why an overthrow of something that really isn't there to begin with, *improves* the structure of science and does not damage it.

Human beings can be divided into two classes: male (M) and female (F). Neither two males by themselves nor two females by themselves can have children (no C). A male and a female together, however, can have children (C). So we can write:

$$M \text{ and } M = \text{no C}$$
$$F \text{ and } F = \text{no C}$$
$$M \text{ and } F = C$$
$$F \text{ and } M = C$$

There is thus the familiar parity situation:
 (1) Like sexes cannot have children.

(2) Opposite sexes can have children.

To be sure, there are sexually immature individuals, barren females, sterile or impotent men, and so on, but these matters are details that don't affect the broad situation. As far as the sexes and children are concerned, we can say that the human species (and, indeed, many other species) conserves parity.

Because the human species conserves sexual parity with respect to childbirth, it is easy to *assume* it conserves it with respect to love and affection as well, so that the feeling arises that sexual love ought to exist only between men and women. The fact is, though, that parity is *not* conserved in that respect and that both male homosexuality and female homosexuality do exist and have always existed. The assumption that parity *ought* to be conserved where, in actual fact, it isn't, has caused many people to find homosexuality immoral, perverse, abhorrent, and has created oceans of woe throughout history.

Again, in Judeo-Christian culture, the institution of marriage is closely associated with childbirth and therefore strictly observes the law of conservation of parity that holds for childbirth. A marriage can take place only between one man and one woman because, ideally, that is the simplest system that makes childbirth possible.

Now, however, there is an increasing understanding that parity, which is rigidly conserved with respect to childbirth, is not necessarily conserved with respect to sexual relations. Increasingly, homosexuality is treated not as a sin or a crime, but as, at most, a misfortune (if that).

There is the further attitude, slowly growing in our society, that there is no need to force the institution of marriage into the tight grip of parity conservation. We hear, more and more frequently, of homosexual marriages and of group marriages. (The old-fashioned institution of polygamy is an example of one kind of marriage, enjoyed by many of the esteemed men of the Old Testament, in which sexual parity was not conserved.)

In the next chapter, then, we'll go on with the nature of the experiment that established the non-conservation of parity in the weak nuclear interaction and consider what happened afterward.

## 2 – THE LEFT HAND OF THE ELECTRON

I received a letter yesterday which criticized my writing style. It complained, 'you avoid the poetic to the extent that when a cryptic, glowing, "charged" phrase occurs to you, I'd be willing to bet that you deliberately put it aside and opt for a clearer but more pedestrian one.'

All I can say to that is that you bet your sweet life I do.

As all who read my volumes of science essays must surely be aware, I have a dislike for the mystical approach to the universe, whether in the name of science, philosophy, or religion. I also have a dislike for the mystical approach to literature.

I dare say it is possible to evoke an emotional reaction through a 'cryptic, glowing, "charged" phrase' but you show me a cryptic phrase and I'll show you any number of readers who, not knowing what it means but afraid to admit their ignorance, will say, 'My, isn't that poetic and emotionally effective.'

Maybe it is, and maybe it isn't; but a vast number of literary incompetents get by on the intellectual insecurity of their readers, and a vast number of hacks write a vast quantity of bad 'poetry' and make a living at it.

For myself, I manage to retain a certain amount of intellectual security. When I read a book that is intended (presumably) for the general public and find that I can make neither head nor tail of it, it never occurs to me that this is because I am lacking in intelligence. Rather, I reach the calmly assured opinion that the author is either a poor writer, a confused thinker, or, most likely, both.

Holding these views, it is not surprising that I 'opt for a clearer but more pedestrian' style in my own writing.

For one thing, my business and my *passion* (even in my fiction writing) is to explain. Partly it is the missionary instinct that makes me yearn to make my readers see and understand the universe as I see and understand it, so that they may enjoy it as I do. Partly, also, I do it because the effort to put things on paper clearly enough to make the reader understand, makes it possible for *me* to understand, too.

I try to teach because whether or not I succeed in teaching others, I invariably succeed in teaching myself.

Yet I must admit that sometimes this self-imposed task of mine is harder than other times. Continuing the exposition on parity and related topics begun in Chapter 1 is one of the harder times, but then no one ever promised me a rose garden, so let's continue—

The conservation laws are the basic generalizations of physics and of the physics aspects of all other sciences. In general, a conservation law says that some particular overall measured property of a closed system (one that is not interacting with any other part of the universe) remains constant regardless of any changes taking place within the system. For instance, the total quantity of energy within a closed system is always the same regardless of changes within the system and this is called 'the law of conservation of energy'.

The law of conservation of energy is a great convenience to physicists and is probably the most important single conservation law, and therefore the most important single law of any kind in all of science. Yet it does not seem to carry a note of overwhelming necessity about it.

*Why* should energy be conserved? Why shouldn't the energy of a closed system increase now and then, or decrease?

Actually, we can't think of a reason, if we think of energy only. We simply have to accept the law as fitting observation.

The conservation laws, however, seem to be connected with symmetries in the universe. It can be shown, for instance, that if one assumes time to be symmetrical, one must expect energy to be conserved. That time is symmetrical means that any portion of it is like any other and that the laws of nature therefore display 'invariance with time' and are the same at any time.

In a rough and ready way, this has always been assumed by mankind – for closed systems. If a certain procedure lights a fire or smelts copper ore or raises bread dough on one day, the same procedure should also work the next day or the next year under similar conditions. If it doesn't, the assumption is that you no longer have a closed system. There may be interference from the outside in the form (mystics would say) of a malicious witch or an evil spirit, or in the form (rationalists would say) of unexpected moisture in the wood, impurities in the ore, or coolness in the oven.

If we avoid complications by considering the simplest possible forms of matter – subatomic particles moving in response to the various fields produced by themselves and their neighbors – we readily assume that they will obey the same laws at any moment in time. If a system of subatomic particles were to be transferred by some time machine to a point in time a century ago or a million years ago, or a million years in the future, the change in time could not be detected by studying the behaviour of the subatomic particles *only*. And if that is true, the law of conservation of energy is true.

Of course, invariance with time is just as much an assumption as the conservation of energy is, and assumptions may not square with observation. Thus, some theoretical physicists have speculated that the gravitational interaction may be weakening in intensity very slowly with time. In that case, you could tell an abrupt change in time by noting (in theory) an abrupt change in the strength of the gravitational field produced by the particles being studied. Such a change in gravitational intensity with time has not yet been actually demonstrated, but if it existed, the law of conservation of energy would be not quite true.

Putting that possibility to one side, we end with two equivalent assumptions: (1) energy is conserved in a closed system, and (2) the laws of nature are invariant with time.

Either both statements are correct or both are incorrect, but it is the second, it seems to me, that seems more intuitively necessary to us. We might not be bothered by having a little energy created or destroyed now and then, but we would somehow feel very uncomfortable with a universe in which the laws of nature changed from day to day.

Consider, next, the law of conservation of momentum. The total momentum (mass times velocity) of a closed system does not vary with changes within the system. It is the conservation of momentum that allows billiard sharps to work with mathematical precision. (There is also an independent law of conservation of angular momentum, where circular movement about some point or line is considered.)

Both conservation laws, that of momentum and that of angular momentum, depend on the fact that the laws of nature are invariant with position in space. In other words, if a group of subatomic particles is instantaneously shifted from here to the neighborhood of Mars, or of a distant galaxy, you could not tell by observing the subatomic particles *alone* that such a shift had taken place. (Actually, the gravitational intensity due to neighboring masses of matter would very likely be different, but we are dealing with the ideal situation of fields originating only with the particles within the closed system, so we ignore outside gravitation.)

Again, the necessity of invariance with space is more easily accepted than the necessity of the conservation of momentum or of angular momentum.

Most other conservation laws also involve invariances of this sort, but not of anything that can be reduced to such easily intuitive concepts as the symmetry of space and time. —Parity is an exception.

In 1927, the Hungarian physicist Eugene P. Wigner showed that conservation of parity is equivalent to right-left symmetry.

This means that for parity to be conserved there must be

no reason to prefer the right direction to the left or vice versa in considering the laws of nature. If one billiard ball hits another to the right of center and bounces off to the right, it will bounce off to the left in just the same way if it hits the other ball to the left of center.

If a ball bouncing off to the right is reflected in a mirror that is held parallel to the original line of travel, the moving ball in the mirror seems to bounce off to the left. If you were shown diagrams of the movement of the real ball and of the movement of the mirror-image ball, you could not tell from the diagrams alone, which was real and which the image. Both would be following the laws of nature perfectly well.

If a billiard ball is itself perfectly spherical and unmarked it would show left-right symmetry. That is, its image would also be perfectly spherical and unmarked, and if you were shown a photograph of both the ball itself and the image, you couldn't tell which was which from the appearance alone. Of course, if the billiard ball had some asymmetric marking on it, like the number 7, you could tell which was real and which was the image, because the number 7 would be 'backward' on the image.

The trickiness of the mirror-image business is confused because we ourselves are asymmetric. Not only are certain inner organs (the liver, stomach, spleen, and pancreas) to one side or the other of the central plane, but some perfectly visible parts (the part in the hair, as an example, or certain skin markings) are also. This means we can easily tell whether a picture of ourselves (or some other familiar individual) is of us as we are or of a mirror image by noting that the part in the hair is on the 'wrong side', for instance.

This gives us the illusion that telling left from right is an easy thing, when actually it isn't. Suppose you had to identify left and right to some stranger where the human body could not be used as reference, to a Martian who couldn't see you, for instance. You might do it by reference to the Earth itself, if the Martian could make out its surface, for the continental configurations are asymmetric, but what if you were talking with someone far out near Alpha Centauri.

The situation is more straightforward if we consider subatomic particles and assume them (barring information to the contrary) to be left-right symmetric, like perfectly spherical unmarked billiard balls. In that case if you took a photograph of the particle and of its mirror image, you could not tell from the appearance alone which was particle and which mirror image.

If the particle were doing something toward our left, then the mirror image would be doing the equivalent toward our right. If, however, both the leftward act and the rightward act were equally possible by the laws of nature, you still couldn't tell which was particle and which was mirror image. —And that is precisely the situation that prevails when the law of conservation of parity holds true.

But what if the law of conservation of parity is not true under certain conditions. Under those conditions, then, the particle is asymmetric or is working asymmetrically; that is, doing something leftward which can't be done rightward, or vice versa. In that case, you can say, 'This is the particle and this is the image. I can tell because the image is backward (or because the image is doing something which is impossible).'

This is equivalent to recognizing that a representation of a friend of ours is actually a mirror image because his hair part is on the wrong side or because he seems to be writing fluently with his left hand when you know he is actually right-handed.

When Lee and Yang (see Chapter 1) suggested that the law of conservation of parity didn't hold in weak nuclear interactions, that meant one ought to be able to differentiate between a weak nuclear event and its mirror image. —And one common weak nuclear event is the emission of an electron by an atomic nucleus.

The atomic nucleus can be considered as a spinning particle, which is symmetrical east and west and also north and south (just as the Earth is). If we take the mirror image of the particle (the 'image-particle'), it seems to be spinning in the 'wrong direction', but are you sure? If you turn the image-particle upside down, it is then spinning in the right direction and it still looks just like the particle. You can't differentiate between the particle and the image-particle by the direction of its spin because you can't tell whether the particle or the image-particle is right side up or upside down. As far as spin is concerned, an upside-down image-particle looks just like a right-side-up particle.

Of course, a spinning particle has two poles, a north pole and a south pole, and to all appearances we can tell which is which. By lining the particle up with a strong magnetic field we can compare the direction of the particle's axis of rotation with that of the Earth and identify the north and south pole. In that way we could tell whether the particle was right side up or upside down.

Ah, but we are using the Earth as a reference here and the Earth is asymmetric thanks to the position and shape of the continents. If we didn't use the Earth as reference (and we shouldn't because we ought to be able to work out the behavior of subatomic particles in deep space far from the Earth) there would be no way of telling north pole from south pole. Whether we considered spin or poles, we couldn't tell a symmetrical particle from its mirror image.

But suppose the particle gives off an electron. Such an electron tends to fly off from one of the poles, but from which? Suppose it could fly off from either pole with equal ease. In that case, if we were dealing with a trillion nuclei giving off a trillion electrons, half would fly off one pole and half off the other. We could not distinguish one pole from the other and we still couldn't distinguish the particle from the image-particle.

On the other hand, if the electrons tended to come off from one pole more often than from the other, we would have a marker for one of the poles. We could say, 'Viewing the particle from a point above the pole that gives off the electrons, it rotates counter-clockwise. That means that this other particle is actually an image-particle, because viewed in that manner it rotates clockwise.'

This is exactly what should be true if the law of conservation of parity does not hold in the case of electron emission by nuclei.

But is it true? When atomic nuclei (trillions of them) are shooting off electrons, the electrons come off in every direction equally – but that is only because the nuclear poles are facing in every direction, in which case electrons would shoot off in all ways alike whether they were coming from one pole only or from both poles equally.

In order to check whether the electrons are coming from both poles or from one pole only, the nuclei must be lined up so that all the north poles are pointing in the same direction. To do this, the nuclei must be lined up by a powerful magnetic field and must be cooled to nearly absolute zero so that they have no energy that will vibrate them out of line.

After Lee and Yang made their suggestion, Madame Chien-Shiung Wu, a fellow physics professor at Columbia University, performed exactly this experiment. Cobalt-60 nuclei, lined up appropriately, shot electrons off the south pole, not the north pole.

In this way, it was proven that the law of conservation of parity did not hold for weak nuclear interactions. This meant one could distinguish between left and right in such cases, and the electron, when involved in weak nuclear interactions, tended to act leftward rather than rightward, so that it can be said to be 'left-handed'.

The electron, which carries a unit negative electric charge, has another kind of 'image'. There is a particle exactly like the electron, but with a unit positive electric charge. It is the 'positron'.

Indeed every charged particle has a twin with an opposite charge, an 'antiparticle'. There is a mathematical operation which converts the expression that describes a particle into one that describes the equivalent antiparticle (or vice versa). This operation is called 'charge conjugation'.

As it happens, if a particle is left-handed, its antiparticle is right-handed, and vice versa.

Observe then, that if an electron is doing something left-handedly, its mirror image would seem to be an electron doing it right-handedly, which is impossible – and the impossibility would serve to distinguish the image from the particle.

On the other hand, if you employed the charge conjugation operation, you would change a left-handed electron into a left-handed positron. The latter is also impossible and this impossibility would serve to distinguish the image from the particle.

In weak nuclear interactions, then, not only does the law of conservation of parity break down, but also the law of conservation of charge conjugation.*

However, suppose you not only alter the right-left of the electron by imagining its mirror image, but also imagine that at the same time you have altered the charge from negative to positive. You have effected both a parity change and a charge conjugation change. The result of this double shift would be the conversion of a left-handed electron into a right-handed positron. Since left-handed electrons and right-handed positrons are both possible, you cannot tell by simply looking at a diagram of each, which is the original particle and which the image.

In other words, although neither parity nor charge conjugation is conserved in weak nuclear interactions, the combination of the two *is* conserved. Using abbreviations we say that there is neither P conservation nor C conservation in weak nuclear interactions, but there is, however, CP conservation.

It may not be clear to you how it is possible for two items to be individually not conserved, yet to be conserved together. Or (to put it in equivalent fashion) you may not see how two objects, each easily distinguishable from its mirror image, are no longer so distinguishable if taken together.

* Both conservation laws are true in strong nuclear interactions, however. In strong nuclear interactions, not only are leftward and rightward equally natural at all times, but anything a charged particle can do, the oppositely charged antiparticle can also do.

Well, then, consider—

The letter *b*, reflected in the mirror is *d*. The letter *d*, reflected in the mirror is *b*. Thus, both *b* and *d* are easily distinguished from their mirror images.

On the other hand, if the combination *bd* is reflected in a mirror, the image is also *bd*. Both *b* and *d* are individually inverted and the order in which they occur is inverted, too. All the inversions cancel and the net result is that although *b* and *d* are altered by reflection, the combination *bd* is not. (Try it yourself with printed lower-case letters and a mirror.)

Let's point out one more thing about left-right reflection. Suppose the solar system were reflected in a mirror. If we observed the image, we would see that all the planets were circling the Sun the 'wrong way' and that the Moon was circling the Earth the 'wrong way', and that the Sun and all the planets were rotating on their axes the 'wrong way'.

If you ignored the asymmetry of the surface structure of the planets, and just considered each world in the solar system to be a featureless sphere, then you could not tell the image from the real thing from their motions alone. The fact that everything was turning the 'wrong way' means nothing, for if you observe the image while standing on your head, then everything is turning the 'right way' again, and in outer space there is no way of distinguishing between standing 'upright' and standing 'on your head'.

And certainly the gravitational interaction, which is the predominant factor in the solar system's working, is unaffected by the reversal of right and left. If all the revolutions and rotations in the solar system were suddenly reversed, gravitational interactions would account for the reversed motions as adequately and as neatly as for the originals.

But consider this—

Suppose that we didn't use a mirror at all. Imagine, instead, that the direction of time reversed itself. The result would be like that of running a movie film backward. With time reversed, the Earth would seem to be going 'backward' about the Sun. All the planets would seem to be going 'backward' about the Sun, and the Moon to be going 'backward' about the Earth. All the bodies of the solar system would be spinning 'backward' about their axis.

But notice that the 'backward' that takes place on reversing time, is just the same as the 'wrong way' that takes place in the mirror image. Reversing the direction of time flow and mirror-imaging space produce the same effect. And there is no way of telling from observing the motions of the solar system alone whether time is flowing forward or backward. This inability to tell the direction of time flow is also true in the case of subatomic reactions (T conservation).*

Or consider this—

An electron moving through a magnetic field pointing in a particular direction will veer to the right. The positron, with an opposite charge, would, when moving in the same direction through the same magnetic field, veer to the left. The two motions are mirror images, so that in this case the shift from a charge to its opposite also produces the same effect as a left-right shift.

Or suppose we reverse the direction of time flow. An electron moving through a magnetic field may veer to its right, but if a picture is taken of the motion and the film is reversed and projected, the electron will seem to be moving backward and, in doing so, will veer to its left. Again, time flow and left-right symmetry are connected.

It would seem then that charge conjugation (C), parity (P), and time reversal (T) are all rather closely related and all somehow connected with left-right symmetry. If, then, left-right symmetry breaks down in weak nuclear interaction with respect to one of these, the symmetry can be restored with one or both others.

If a particle is doing something leftward, and its image is

* We can tell the direction of time flow under ordinary conditions easily enough because of entropy-change effects. This produces the equivalent of an asymmetry in time. Where entropy change is zero, however, as in planetary motions and subatomic events, T is conserved.

doing something rightward, which is impossible (so that the image can be spotted through a breakdown in P conservation), you can reverse the charge on the image-particle and convert the action into a possibility. If the action is impossible even with the reversed charge (so that the image can be spotted through a breakdown in CP conservation), you can reverse the direction of time flow, and then you will find the action *is* possible. In other words, there is 'CPT conservation' in the weak nuclear interaction.*

The result is that the universe is symmetrical, as it has always been thought to be, with respect to strong nuclear interactions, electromagnetic interactions and gravitational interactions.

Only weak nuclear interactions have been in question and there the failure of the law of conservation of parity seemed to introduce a basic asymmetry to the universe. The broadening of the concept to CPT conservation restored the symmetry – but only in theory.

Does CPT conservation actually present us with a symmetrical universe in practice? As far as P (parity) is concerned, there is an equal supply of rightness and leftness in the universe. As far as T (time reversal) is concerned, there is also an equal supply of pastness and futureness. But where C (charge conjugation) is concerned, symmetry in practice breaks down.

The most common subatomic particles to be involved in weak nuclear interactions are the electron and the neutrino. For symmetry to exist in practice, then, there should be equal supplies of electrons and positrons and equal supplies of neutrinos and antineutrinos. This, however, is not so.

Certainly on Earth, almost certainly throughout our Galaxy, and, for anything we know to the contrary,

* Actually, there was some indication in recent years that CPT is *not* invariably conserved in weak nuclear interactions and physicists have been examining the possible consequences in rather perturbed fashion. However, all the returns don't seem to be in here and we'll have to wait and see.

throughout the entire universe, there are vast numbers of electrons and neutrinos, and hardly any positrons and antineutrinos.

The universe then – at least our universe – or at the very least our section of our universe – is electronically left-handed and that may have had an interesting effect on the development of life.

In order to explain that, I must change the subject radically, however, and make a new start. That I will do in the next chapter.

## 3 – SEEING DOUBLE

I currently do my writing in a two-room suite in a hotel, and about a month ago I became aware of someone banging loudly against the wall in the corridor outside. Naturally, I was furious. Did whoever it was not realize that within my rooms the most delicate work of artistic creation was going on?

I stepped into the corridor and there, on a ladder, at the elevators, was an honest workingman banging a hole into the wall for some arcane purpose of his own.

'Sir,' I said, with frowning courtesy, 'how long do you intend to make the world hideous with your banging at that hole?'

And the horny-handed son of toil turned his sweat-streaked face in my direction and answered jauntily, 'How long did it take Michelangelo to do the ceiling in the Sistine Chapel?'

What could I do? I burst out laughing, went back to my cell, and worked cheerfully along to the tune of banging which I no longer resented since it was produced by an artist who knew his own worth.

Things take as long as they take, in other words. And even Michelangelo's long stint on his back, painting that fresco, pales into insignificance in comparison to the length of the intervals it took to build some corner or other of the majestic structure of science.

In the seventeenth century, for instance, a question arose about light which wasn't answered for 148 years, despite the fact that, till it was answered, no theory as to the nature of light could possibly hold water.

The story begins with Isaac Newton, who, in 1666, passed a beam of sunlight through a prism and found that the beam was spread out into a rainbowlike band which he called a 'spectrum'.

Newton felt that since light traveled in a straight path, it must be made up of a stream of very fine particles, moving at an enormous speed. These particles differed among themselves in some way so that they produced the sensations of different colors. In sunlight, all the different particles were mixed evenly and the effect was to impress our eye as white light.

In passing obliquely into glass, however, the light particles bent sharply in their path; that is, they were 'refracted'. Particles differing in their color nature were refracted by different amounts so that the colors in white light were separated within the glass. In an ordinary sheet of glass, with two parallel faces, the effect was reversed when the light emerged once more from the other side, so that the colors were again merged into white light.

In a prism, it was different. The light particles bent sharply when they entered one side of the triangular piece of glass, and then bent a second time in the *same* direction on emerging through a second, non-parallel side. The colors, separated on entering the prism, were even farther separated on emerging.

All this made excellent sense, and Newton backed it up with careful experimentation and reason. And yet exactly what was different about the particles that gave rise to the various colors, Newton couldn't say.

His contemporary the Dutch physicist Christiaan Huygens suggested in 1678 that light was a wave phenomenon. This made it possible to explain the different colors easily. A light wave would have to have some particular length, and light of different wavelengths might well impress the eyes as of different colors (just as sound of different wavelengths impresses the ears as of different pitch).

Still, waves had their own problems. All man's experience with waves (water waves, for instance, and sound waves) made it clear that waves curved around obstructions. Light on the other hand traveled in a straight line past obstructions and cast sharp shadows.

Huygens tried to explain that away by presenting a mathematical line of reasoning that showed that the ability to curve about an obstruction depended on the length of the wave. If light waves were *much* shorter than sound waves or water waves, they would then not bend, detectably, about ordinary obstructions.

Newton recognized the convenience of the wave theory, but could not go along with the suggestion of waves so tiny they would cast sharp shadows. He stuck to particles and such was his eventual prestige that scientists, by and large, went along with the particle theory of light in order not to place themselves in disagreement with Newton.

But in 1669, a Danish physician, Erasmus Bartholinus – a thoroughly obscure individual – made an observation which assured him a place in the history of science, for it raised a question the giants could not answer.

Bartholinus had received a transparent crystal which had been obtained in Iceland, so that it became known as 'Iceland spar', where 'spar' is an old-fashioned term for a non-metallic mineral.*

The crystal was shaped like a rhombohedron (a kind of slanted cube), with six flat faces, each one parallel to the one on the opposite side. Bartholinus was studying the properties of this crystal and I presume he placed it on a piece of paper with writing or printing on it, on one occasion. When he picked it up, he noticed that the writing or printing was double when viewed through the crystal.

In fact, when one looked through the crystal, it turned out one was seeing double. Apparently each beam of light entering the crystal was refracted, but not all to the same extent. Part of the light was refracted a certain amount and the remainder another and greater amount, so that though one beam entered the crystal, two beams emerged.

* Actually, Iceland spar is a transparent variety of calcium carbonate, if that helps any.

The phenomenon was called 'double refraction'.

Any theory of light had to explain double refraction, and neither Huygens nor Newton could do so. Apparently, the waves, or particles, of light must fall into two sharply defined classes so that one class can behave in one way and the other class in another. The two-way difference can have nothing to do with color, for all colors of light were equally double-refracted by Iceland spar.

Huygens' view of light waves was that they were 'longitudinal waves'; that is, similar to sound waves in structure (though much shorter in length) and that they represented a series of compressions and rarefactions in the ether they passed through. Huygens did not see how such longitudinal waves could fall into two sharply different classes.

Nor could Newton see how light particles could be divided into two sharp classes. He speculated, rather vaguely, that the particles might differ among themselves in some fashion analogous to the two opposed poles of a magnet, but he didn't follow that up, since he was at a loss for any way of finding evidence for the suggestion.

Physicists were forced to back away. Bartholinus' observation didn't fit either of the current theories of light, so, as far as possible, it was to be ignored.

This was not wickedness on the part of scientists; nor the obtuse workings of a conspiratorial 'establishment'. On the contrary, it makes sense.

Suppose a piece doesn't seem to fit a jigsaw puzzle. If you stop everything and start worrying exclusively about that troublesome piece, you may never get anywhere. If, however, you ignore the piece and continue working at other parts of the jigsaw, using whatever system seems convenient, you may eventually reach a point where, through the *other* work, new understandings are reached, and suddenly the old piece that was once so troublesome fits into place with no trouble at all.

Double refraction was not forgotten altogether, of course. Even as late as 1808, it was still sticking in the scientific gizzard, and the Paris Academy offered a prize for the best mathematical treatment of the subject. A twenty-three-year-old French army engineer, named Etienne Louis Malus (who accepted Newton's particle theory) decided to see what he could do in that direction. He got some doubly refracting crystals and began to experiment with them. As it happened, he did not win the prize, but he made an interesting observation and coined a phrase that entered the scientific vocabulary.

From his room he could see out on the Luxembourg Palace and, at one time, sunlight was reflected from a window of that palace into his room. Idly, Malus pointed a doubly refracting crystal in that direction, expecting to look through it and see two windows. He did not! He saw only one window.

Apparently what happened was that the window, in reflecting the sunlight, reflected only one of the two classes of light particles.

Malus remembered that Newton had said that the light particle varieties might be analogous to the opposing poles of a magnet. Thinking along those lines, he felt that only one pole of light had been reflected, and that the beam shining into his room contained only particles with that one pole.

Malus therefore spoke of the light beam that entered his room as consisting of 'polarized light'. The phrase stands to this day, even though it is based on a false speculation, and even though the notion of poles of light was, in actual fact, being killed dead even before Malus had made his observation.

In 1801, you see, an English physician, Thomas Young, began a series of experiments in which he showed that one beam of light could somehow cancel another intermittently, so that the two would not combine to give a smooth field of light, but rather a series of bands, alternately light and dark.

If light consisted of particles, such 'interference' was extremely difficult to explain. How could one particle cancel another?

If light consisted of waves, however, interference was childishly easy to explain. If light consisted of alternate rarefactions and compressions, for instance, then if two light beams fitted together so that the compressed area in one beam fell on the rarefied area in the other and vice versa, the two lights would indeed cancel out into darkness.

Young was able to explain every characteristic of his interference pattern by Huygens' wave theory. To be sure, many physicists (especially English physicists) tried to object, in the name of Newton. However, not even the most glorious name can long resist observations that anyone can confirm and explanations that explain perfectly. —So the wave theory won out.

Yet Young could not explain double refraction any better than Huygens had.

But then, in 1817, a French physicist, Augustin Jean Fresnel, suggested that perhaps the light waves were not longitudinal after the fashion of sound waves, and did not represent alternate compressions and rarefactions in the ether. Perhaps, instead, they were 'transverse waves', like those on water surfaces; waves which moved up and down at right angles to the line of propagation of the wave.

Transverse waves could explain interference just as well as longitudinal waves did. If two light beams merged, and one was waving up where the other was waving down, and vice versa, the two would cancel, and two lights would make darkness.

Water waves, which serve as a model for light waves, can only move up and down at right angles to the two-dimensional water surface. A ray of light, however, has greater freedom. Imagine such a ray moving toward you. It could wave up and down, or right and left, or anything in between and always be waving at right angles to the direction in which it was moving. (You can see what this means concretely, if you tie one end of a long rope to a post and make waves in it, up and down, right and left, or obliquely.)

Once such transverse waves were proposed, they were accepted with remarkably little trouble, for through them, the phenomenon of double refraction could finally be explained, 148 years after the problem had arisen.

To see that, consider that the light waves in an ordinary beam of light could be waving in all possible directions at right angles to the direction of travel – up and down, left and right, and all degrees of in-betweenness. That would represent ordinary or 'unpolarized' light.

Suppose, though, there were some way of dividing the light into two varieties, one in which all the waves move up and down, and the other in which all the waves move left and right.

For each wave in unpolarized light which vibrates obliquely, there would be a division into two waves, of lesser energy, of the permitted classes.

If a particular wave were just at forty-five degrees to the vertical, just halfway between the up and down and the left and right, it would be divided into two waves, one up and down and the other left and right, each with half the energy of the original. If the oblique wave were nearer horizontal than vertical, then it would be broken up into two waves, with the left and right having the greater supply of energy. If it were nearer the vertical, then the up and down would end with the greater supply of energy.

It is easy to show, in fact, that a beam of unpolarized light can be divided into two beams of equal energy, in one of which all the transverse waves are in one direction, while in the other all the transverse waves are in a plane at right angles to the first. Since in each case all the waves move in a single plane, the unpolarized beam of light can be viewed as broken up into two mutually perpendicular 'plane-polarized' beams.

But what causes light to break up into plane-polarized beams? Certain crystals do. Crystals are made up of serried ranks and files of atoms arranged in very orderly array. Light, in passing through, is sometimes compelled to take up waves in certain planes only.

(You can see a crude analogy of this if you pass a rope

through a picket fence and tie it to a pole somewhere on the other side. If you make up-and-down waves in the rope, they will pass through the opening between the pickets, so that the rope on the other side of the fence also waves. If you make waves left and right, the pickets on either side of the opening stop those waves and the rope on the other side of the pickets does not wave. If you make the rope wave in every which way, only those waves which will fit between the pickets at least partly will get through, and on the other side of the fence, whatever you do, there will only be up-and-down waves. The picket fence polarizes the 'rope waves'.)

Crystals such as Iceland spar will permit only two planes of vibration, one perpendicular to the other. Unpolarized light entering Iceland spar breaks up into two mutually perpendicular plane-polarized beams within. The two beams of polarized light interact differently with the atoms, travel at different velocities and the slower beam is refracted through a greater angle. The two beams take separate paths within the crystal and emerge in difference places. It is for that reason that looking through Iceland spar causes you to see double, and Bartholinus' puzzle is solved.

Plane polarization can also take place on reflection. If an unpolarized beam strikes a reflecting surface at an angle, it often happens that those particular waves which occupy a certain plane are more efficiently reflected than those in other planes. The reflected beam is then heavily or even entirely plane-polarized and Malus' puzzle is solved.

In 1828, a Scottish physicist, William Nicol, introduced a new refinement to Iceland spar. He sawed a crystal in half in a certain fashion* and cemented the halves together with Canadian balsam. When light enters the crystal, it splits up into two plane-polarized beams, which travel in slightly different directions and hit the Canadian balsam seam at slightly different angles. The one that hits it at the lesser

* I am tempted every once in a while to present diagrams, and on rare occasions I do. I am, however, primarily a word-man and I try not to lean on pictorial crutches. In this case, the exact manner of dividing the crystal doesn't affect the argument, so the heck with it.

## SEEING DOUBLE

angle to the perpendicular passes through into the other half of the crystal and eventually emerges into open air. The one that hits it at the greater angle is reflected and never enters the other half of the crystal.

In other words, a beam of unpolarized light enters the 'Nicol prism' at one end and a single beam of plane-polarized light emerges at the other end.

Now imagine two Nicol prisms lined up in such a way that a beam of light passing through one will continue on into the second. If the two Nicol prisms are lined up parallel, that is, with the atom arrangements oriented in identical fashion in both, the beam of polarized light emerging from the first passes also through the second without trouble.

(It is like a rope passing through two picket fences in both of which the pickets are up and down. An up-and-down rope wave that passes between the pickets in the first fence will also pass between the pickets in the second.)

But what if the two Nicol prisms are oriented perpendicularly to each other? The plane-polarized beam emerging from the first Nicol prism is refracted through a greater angle by the second one and is reflected from the Canadian balsam seam in it. No light at all emerges from the second prism. (If we go back to the picket fence analogy, and have the pickets in the second fence arranged horizontally, you will see that any up-and-down waves that get through the first fence will be stopped by the second. No rope waves of any kind can go through two fences in one of which the pickets are vertical and in the other horizontal.)

Suppose, next, that you arrange to have the first Nicol prism fixed in place, but allow the second Nicol prism to be rotated freely. Arrange also an eyepiece through which you can look and see the light that passes through both Nicol prisms.

Begin with the two Nicol prisms arranged in parallel fashion. You will see a bright light in the eyepiece. Slowly rotate the second prism, which is nearer your eye. Less and less of the light emerging from the first prism can get through the second, since more and more of it is reflected at the second's Canadian balsam seam. The light you see becomes dimmer and dimmer as you rotate the second prism, until, when you have turned through ninety degrees, you see no light at all. The same thing happens whether you rotate the prism clockwise or counterclockwise.

Using such a pair of Nicol prisms you can determine the plane of vibration of a beam of polarized light. Suppose such a beam emerges from the fixed Nicol prism, but you are not sure as to exactly how that prism is oriented. That means you don't know the location of the plane of vibration of the light emerging. In that case, you need only turn the rotating Nicol prism until the beam of light you see through it is at its brightest.* At that point, the second prism is oriented parallel to the first and from the position of the second you know the plane of vibration of the polarized light.

For this reason the first, fixed, Nicol prism is called the 'polarizer', and the second, rotating, one, the 'analyzer'.

Now imagine an instrument in which there is a space between polarizer and analyzer into which a standard tube can be placed containing some liquid transparent to light. To make sure conditions are always the same, the temperature is kept at a fixed level, light of a single fixed wavelength is used, and so on.

If the tube contains distilled water, nothing happens to the plane of polarized light emerging from the polarizer. The air, the glass, the water all may and do absorb a trifle of light, but the analyzer continues to mark the plane at the same point. If a salt solution is used in place of distilled water, the same thing is true.

But place sugar solution in the tube, and something new happens. The light you see through the analyzer is now greatly dimmed and this is not the result of absorption.

---

* It isn't so easy to tell when the light is brightest, but there is a device whereby the circle of light you see is divided into two half-circles and you turn the prism until the two half-circles are equally bright, something easy to determine.

## SEEING DOUBLE

Sugar solution doesn't absorb light significantly more than water itself does.

Besides, if you rotate the analyzer, the light brightens again. You can eventually get it as bright as it was originally, provided you completely alter the orientation of the analyzer. What it amounts to is that the sugar solution has rotated the plane of polarized light. Anything which does this is said to display 'optical activity'. The instrument used to detect optical activity and measure its extent is called a 'polarimeter'.

A useful polarimeter was first devised in 1840 by the French physicist Jean Baptiste Biot. He had pioneered in the study of optical activity long before he devised the polarimeter (to make his work easier and more precise) and even before Nicol had first constructed his prism.

As early as 1813, for instance, Biot reported certain observations that were eventually interpreted according to the new transverse-wave theory. It turned out that a quartz crystal, correctly cut, rotated the plane of polarized light passing through it. What's more, the thicker the piece of quartz, the greater the angle through which the plane was rotated. And still further, some pieces of quartz rotated the plane clockwise and some rotated it counterclockwise.

The usual way of reporting the clockwise rotation was to say that the plane of polarization had been rotated to the right. Actually, this is a careless and ambiguous way of reporting it. If the plane is viewed as straight up and down, then the upper end of it is indeed rotated to the right when it is twisted clockwise, but the lower end is rotated to the left. Vice versa, in the case of counterclockwise rotation.

However, once a phrase enters the literature it is hard to change no matter how poor, inappropriate, or downright wrong it is. (Look at the phrase 'polarized light' itself, for instance.) Consequently, something that rotates the plane of polarized light clockwise, is said to be 'dextrorotatory' ('right-rotating') and something that rotates it counterclockwise is 'levorotatory' ('left-rotating').

What Biot had shown was that there were two kinds of quartz crystals, dextrorotatory and levorotatory. Using initials, we can speak of *d*-quartz and *l*-quartz.

As it happens, quartz crystals are rather complicated in shape. In certain varieties of those crystals, just those varieties which show optical activity, it can be seen that there are certain small faces that occur on one side of the crystal, but not the other, introducing an asymmetry. What's more, there are two varieties of such crystals, one of which has the odd face on one side, the other of which has it on the other.

The two asymmetric varieties of quartz crystals are mirror images. There is no way in which you can twist one variety through three-dimensional space in order to make it look like the other, any more than you can twist a right shoe so as to make it fit a left foot. And one of these varieties is dextrorotatory, while its mirror image is levorotatory.

It was quite convincing to suppose that an asymmetric crystal will rotate the plane of polarized light. The asymmetry of the crystal must be such that the light beam, traveling through, must be constantly exposed to an asymmetric force, one which pulls, so to speak, more strongly in one direction than the other. So the plane twists and keeps on twisting at a steady rate the greater the distance it must pass through such a crystal. What's more, if a crystal twists the plane of light in one direction, it is inevitable that, all else being equal, the mirror-image crystal will twist the plane in the opposite direction.

You might even argue further that *any* substance which will crystallize in either of two mirror-image forms will be optically active. Furthermore, if two mirror-image crystals are taken of the same substance and of the same thickness, and if all the circumstances are equal (such as temperature and wavelength of light), then the two crystals will show optical activity to precisely the same extent – one clockwise, the other counterclockwise.

And, indeed, all evidence ever gathered shows all of this to be perfectly correct.

But then, Biot went on to spoil the whole thing by discovering that certain liquids, such as turpentine, and certain solutions, such as camphor in alcohol and sugar in water, are also optically active.

This presents a problem. Optical activity is tied in firmly with asymmetry in all work on crystals, but where is the asymmetry in the liquid state? None that any chemist could see in 1840.

Once again, then, the solution of one problem in science served to raise another. (And thank heaven for that, or where would there be any interest in science?) Having solved Bartholinus' problem and Malus' problem by establishing the existence of transverse light waves, science found itself with Biot's problem – how a liquid which seemed to have no asymmetry about it could produce an effect that seemed to be logically produced only by asymmetry.

Which brings us to Louis Pasteur's first great adventure in science – next chapter.

## 4 – THE 3-D MOLECULE

In the days when I was actively teaching, full time, at a medical school, there was always the psychological difficulty of facing a sullen audience. The students had come to school to study medicine. They wanted white coats, a stethoscope, a tongue depressor, and a prescription pad.

Instead, they found that for the first two years (at least, as it was in the days when I was actively teaching) they were subjected to the 'basic sciences'. That meant they had to listen to lectures very much in the style of those they had suffered through in college.

Some of those basic sciences had, at least, a clear connection with what they recognized as the doctor business, especially anatomy, where they had all the fun of slicing up cadavers. Of all the basic sciences, though, the one that seemed least immediately 'relevant', farthest removed from the game of doctor-and-patient, most abstract, most collegiate, and most saturated with despised Ph.D.'s as professors was biochemistry. —And, of course, it was biochemistry that I taught.

I tried various means of counteracting the natural contempt of medical student for biochemistry. The device which worked best (or, at least, gave me most pleasure) was to launch into a spirited account of 'the greatest single discovery in all the history of medicine' – that is, the germ theory of disease. I can get very dramatic when pushed, and I would build up the discovery and its consequences to the loftiest possible pinnacle.

And then I would say, 'But, of course, as you probably all take for granted, no mere physician could so fundamentally

## THE 3-D MOLECULE

revolutionize medicine. The discoverer was Louis Pasteur, Ph.D., a biochemist.'

Pasteur's first great discovery, however, had nothing to do with medicine, but was a matter of straight chemistry. It involved the matter of optically active substances, a subject I discussed in the previous chapter. To see how he contributed, let's start at the beginning.

In the wine-making process of the fermentation of grape juice, a sludgy substance separates and is called 'tartar', a word of unknown origin. From this substance, the Swedish chemist Karl Wilhelm Scheele in 1769 isolated a compound which had acid properties and which he naturally called 'tartaric acid'.

In itself this wasn't terribly important, but then in 1820, a German manufacturer of chemicals, Charles Kestner, prepared something he felt ought to be tartaric acid and yet didn't seem to be. For one thing, it was distinctly less soluble than tartaric acid. A number of chemists obtained samples and studied it curiously. Eventually, the French chemist Joseph Louis Gay-Lussac named this substance 'racemic acid' from the Latin word for a 'cluster of grapes'.

The more closely racemic acid and tartaric acid were studied, the more puzzling were the differences in properties. Analysis showed that each acid had exactly the same proportion of exactly the same elements in their molecules. Using modern symbols, the formula for each compound was $C_4H_6O_6$.

In the early nineteenth century, when the atomic theory had only been in existence for a quarter of a century or so, chemists had decided that every different molecule had a different atomic content, that it was, in fact, the difference of atomic content that was responsible for the difference of properties. Yet here were two substances, quite distinguishable, with molecules made up of the same proportions of the same elements. It was very disturbing, especially since this was *not* the first time such a thing had been reported.

In 1830, the staunchly conservative Swedish chemist Jöns

Jakob Berzelius,* who didn't believe that molecules with equal structures but different properties were possible, examined both tartaric acid and racemic acid in detail. With considerable chagrin, he decided that even though he didn't believe it, it was nevertheless so. He bowed to the necessary, accepted the finding, and called such equal-structure-different-property compounds 'isomers' from Greek words meaning 'equal proportions' (of elements, that is).

But how could isomers have the same atomic composition and yet be different substances? One possibility is that it is not just the number of atoms of each element that is distinctive, but their physical arrangement within the molecule. This thought, however, was something chemists shuddered away from. The whole notion of atoms was a shaky one. Atoms were useful in explaining chemical properties but they could not be seen or detected in any way and they might very well be no more than convenient fictions. To start talking about actual arrangements within the molecules was to advance farther down the road of accepting atoms as real entities than most chemists cared to – or dared to.

The phenomenon of isomerism was therefore left unaccounted for and kept suspended until such time as chemical advance might produce an explanation.

One difference in properties between tartaric acid and racemic acid was particularly interesting. A solution of tartaric acid or of its salts (that is, compounds in which the acid hydrogen of the compound was replaced by an atom of such elements as sodium or potassium) was optically active. It rotated the plane of polarized light clockwise and was there-

* I have a tendency (as you may occasionally have noticed) to mention large numbers of scientists and to give the contribution of each whenever I get science-historical. This is not a matter of name-dropping. Every advance in science is the result of the co-operative labor of a number of people, and I like to demonstrate that. And I am careful to mention nationalities because it is also important to recognize the fact that science is international in scope.

fore dextrorotatory (see the previous chapter), so that the compound could well be called *d*-tartaric acid.

A solution of racemic acid, on the other hand, was optically inactive. It did *not* rotate the plane of polarized light in either direction. This difference in properties was clearly demonstrated by the French chemist Jean Baptiste Biot, whom I mentioned in the previous chapter as a pioneer in the science of polarimetry.

No one at the time knew why any substance should be optically active in solution, but they did know this— Those crystals known to be optically active had asymmetric structures. In that case, if one were to prepare crystals of tartaric acid and racemic acid or of their respective salts, it would surely turn out that those of the former were asymmetric and those of the latter, symmetric.

In 1844, however, the German chemist Eilhardt Mitscherlich undertook this investigation. He formed crystals of the sodium ammonium salt of both tartaric acid and racemic acid, studied them carefully, and announced that the two substances had absolutely identical crystals.

The basic findings of the budding science of polarimetry were blasted by this report and for the moment all was confusion.

It was at this point that the young French chemist Louis Pasteur entered the scene. He was only in his twenties and his scholastic record at school had been mediocre, yet he had the temerity to suspect it possible that Mitscherlich (a chemist of the first rank) might have been mistaken. After all, the crystals he studied were small and perhaps some tiny details were overlooked.

Pasteur applied himself to the matter and began to produce the crystals and study them painstakingly under a hand lens. He finally decided that there was a definite asymmetry to the crystals of the sodium ammonium salt of tartaric acid. So far, so good. That, at least, was to be expected, since the substance was optically active.

But was it possible now that the sodium ammonium salt of

racemic acid yielded crystals of precisely the same sort, as Mitscherlich maintained? In that case, there would be asymmetric crystals of a substance which was *not* optically active, and that would be very unsettling.

Pasteur produced and studied the crystals of the salt of racemic acid and found that they were indeed also asymmetric *but that not all the crystals were identical*.

Some of the crystals were exactly like those of the sodium ammonium salt of tartaric acid, but others were mirror images of the first group and were asymmetric in the opposite sense.

Could it be that racemic acid was half tartaric acid and half the mirror image of tartaric acid, and that the reason racemic acid was optically inactive was that it was made up of two parts, one part of which neutralized the effect of the other part?

This had to be checked directly. Making use of his hand crystal and a pair of tweezers, Pasteur began to work over those tiny crystals of the racemic acid salt. All those which were right-handed he shoved to one side; all those which were left-handed, to the other. It took him a long time, for he wanted to make no mistake, but he was eventually done.

He then dissolved each set of crystals in a separate sample of water and found both solutions to be optically active!

One of the solutions was dextrorotatory, exactly as tartaric acid was. In fact, it *was* tartaric acid, in every sense.

The other was levorotatory, and differed from tartaric acid in rotating the plane of polarized light in the opposite direction. It was *l*-tartaric acid.

Pasteur's conclusion, announced in 1848, when he was only twenty-six, was that racemic acid was optically inactive only because it consisted of equal quantities of *d*-tartaric acid and *l*-tartaric acid.

The announcement created a sensation and Biot, the grand old man of polarimetry, who was seventy-four years old at the time, cautiously refused to accept Pasteur's finding. Pasteur therefore undertook to demonstrate the matter to him in person.

Biot gave the young man a sample of racemic acid which he had personally tested and which he knew to be optically inactive. Under Biot's shrewd, old eyes, alert for hanky-panky, Pasteur formed the salt, crystallized it, isolated the crystals, and separated them painstakingly by means of hand lens and tweezers. Biot then took over. He personally prepared the solutions from each set of crystals and placed them in the polarimeter.

You guessed it. He found that both solutions were optically active, one in the opposite sense to the other. After that, with typical Gallic enthusiasm, he became fanatically pro-Pasteur.

Actually, Pasteur had been most fortunate. When the sodium ammonium salt of racemic acid crystallizes, it doesn't have to form separate mirror-image crystals. It might also form combination crystals in each of which are equal numbers of molecules of *d*-tartaric acid and *l*-tartaric acid. These combination crystals are symmetrical.

Had Pasteur obtained these crystals he would still have noted their difference from those of the sodium ammonium salt of tartaric acid and have refuted Mitscherlich. On the other hand, he would have missed the far greater discovery of the reason for the optical inactivity of racemic acid and he would also have missed having been the very first man to form optically active substances from an optically inactive start.

As it happens, only symmetric-combination crystals are formed out of solutions above 28°C. (82°F.). It requires solutions of sodium ammonium salt of racemic acid at temperatures below 28°C. to form separate sets of asymmetric crystals. Furthermore, the crystals formed are usually so tiny that they are far too small to separate with hand lens alone. It just happened that Pasteur was working at low temperatures and under conditions which produced fairly good-sized crystals.

Pasteur might be dismissed as an ordinary man who took advantage of an unexpected good break, but (as I used to tell my biochemistry class) he managed to take advantage of similarly unexpected good breaks every five years or so. After a while, you had to come to the conclusion that it was Pasteur who was remarkable and not the laws of chance.

As Pasteur himself once said, 'Chance favors the prepared mind.' We all get our share of lucky breaks and the great man is he who is capable of recognizing a break when it comes, and of taking advantage of it.

Pasteur continued to interest himself in the matter of the tartaric acids. He found that if he heated $d$-tartaric acid for prolonged periods under certain conditions, some of the molecules would change to the $l$-form and racemic acid would be produced. (Ever since, the ability to change optical activity to optical inactivity by heat or by some chemical process through formation of some of the oppositely active form has been known as 'racemization'.)

Pasteur also found a kind of tartaric acid which was optically inactive, which could not be separated into opposite forms under any conditions, and which possessed properties distinct from those of racemic acid. He called it *meso*-tartaric acid, from the Greek word for 'intermediate', since it seemed intermediate between the $d$- and the $l$-forms of the acid.

But all these facts could not explain the existence of optical activity in solutions. Granted that some crystals are symmetrical, while others are asymmetric in one sense or the other, still there are no crystals in solution. There are only molecules.

Could not the molecules themselves retain the asymmetry of the crystals? Was not the asymmetry of the crystals but a reflection of that of the molecules that composed them? Was not racemization a result of the heat-induced rearrangement of atoms within the molecule? Pasteur was sure of all this, but he could think of no way of proving it or of demonstrating what the arrangements must be.

In the 1860's, to be sure, the German chemist Friedrich August Kekule worked out a system whereby a molecule was pictured not merely as a conglomeration of so many atoms

## THE 3-D MOLECULE

of this element or that, but as a collection of atoms connected to one another in a definite arrangement (see Chapter 13). Little dashes were used between symbols of the elements to represent the 'bonds' linking one atom to another, so that the molecule did get to look like a Tinker Toy.

However, the Kekule structures were considered to be highly schematic and to be merely another useful tool for chemists who were working out organic structures and reactions. As in the case of atoms themselves, chemists were not prepared to say that the Kekule structures actually represented the true situation within the molecules.

The Kekule structures did explain the existence of many isomers, since they demontrated gross differences in atomic arrangement even when the total numbers of atoms of each element present within the molecules were the same. The Kekule structures did *not*, however (as used originally), account for those 'optical isomers' which differed *only* in the way in which they twisted the plane of polarized light.

We next come to the Dutch chemist Jacobus Hendricus van't Hoff, who took up the problem in 1874, when he was only twenty-two. The following represents what may have been his line of reasoning.

According to the Kekule system, a carbon atom is represented by the letter *C* with four little bonds attached to it. Usually, these little bonds are shown pointing to the corners of an imaginary square, thus, [C], so that the angle between any two adjacent bonds is ninety degrees. A carbon atom will combine with four hydrogen atoms to form the substance methane, which will then look like this:

$$\begin{array}{c} H \quad\ \ H \\ \diagdown\ \diagup \\ C \\ \diagup\ \diagdown \\ H \quad\ \ H \end{array}$$

Are the four bonds identical? If each is different from the

rest, somehow, then what would happen if one of the hydrogen atoms is replaced by a chlorine atom to form 'methyl chloride'? Surely, there would then be four different methyl chlorides, depending on which of the four different bonds the chlorine atom happened to attach itself to.

But there aren't. There is only one methyl chloride and no more. This indicates that the four carbon bonds are equivalent and, indeed, if the four are drawn to the corners of a square, that is what should be expected. One corner of the square should be no different from any other.

Consider the situation, though, if *two* chlorine atoms replace hydrogen atoms to form 'methylene chloride'. Then, if we still deal with bonds pointing to the corners of a square, there ought to be two different methylene chlorides, depending on whether the two chlorine atoms are placed at adjacent corners of the square or at opposite corners, thus:

$$\begin{array}{ccc} H & & Cl \\ & C & \\ H & & Cl \end{array} \quad \text{OR} \quad \begin{array}{ccc} H & & Cl \\ & C & \\ Cl & & H \end{array}$$

But there aren't. There is only one methylene chloride and no more, which shows that the Kekule structures can't possibly correspond to reality (and, of course, no one claimed that they did).

One way in which they were almost certain not to correspond to reality was that all were drawn, for convenience' sake, in two dimensions – that is, in a plane – and surely it was unlikely that all molecules would be strictly planar in nature.

The four bonds of the carbon atoms were almost certainly distributed in three dimensions and it was only necessary to choose some 3-D arrangement in which each bond was equally adjacent to all three remaining bonds. Only then would there be only a single methylene chloride.

The simplest way of arranging this was to have the four bonds pointing toward the apices of a tetrahedron.* The carbon atom then looks as though it were resting on three bonds forming a squat tripod while the fourth bond is pointing straight up. It doesn't matter which bond you point upward, the other three always form the squat tripod. The carbon atom can thus stand in each of four different positions and look the same each time. What's more, any one bond is equally far from each of the other three. The angle between any two bonds is $109\frac{1}{2}°$.

If we deal with such a 'tetrahedral carbon', then as long as two of the bonds are attached to identical atoms (or groups of atoms), it doesn't matter what atoms, or groups of atoms, are attached to the other two; in every case all possible arrangements are equivalent and only one molecule is formed.

Thus, if attached to the four bonds of a carbon atom are *aaaa*, or *aaab*, or *aabb*, or *aabc*, then it doesn't matter to which bond which atom is attached. If you attach them so as to form what seem to be two different arrangements, then by twisting the first arrangement so that some different bond faces upward, you can make it identical with the second.

Not so when you have four different atoms or groups of atoms attached to the four bonds: *abcd*. In that case, it turns out there are two different and distinct arrangements possible, one of which is the mirror image of the other. No amount of twisting and turning can then make one arrangement look like the other.

A carbon atom to which four different atoms or groups of atoms are attached is an 'asymmetric carbon'.

It turns out that optically active organic substances invariably have asymmetric molecules if the Van't Hoff system is used. Almost always there is at least one asym-

* A tetrahedron is a solid bounded by four equilateral triangles. It can best be understood if it is inspected in the form of a three-dimensional model. Failing that, you are probably familiar with the shape of the Egyptian pyramids – a square base, with each wall slanting inward from one side of that base toward an apex on the top. Well, if you imagine a triangular base instead, you have a tetrahedron.

metric carbon present. (Sometimes there is an asymmetric atom other than carbon present and sometimes the molecule as a whole is asymmetric even though none of the carbon atoms are.)

In tartaric acid there are present two asymmetric carbon atoms. Either can be present in a certain configuration or in its mirror image. Let's refer to these arbitrarily as $p$ and $q$ (since $q$ is the mirror image of $p$). If the two carbon atoms are $pp$, then we have $d$-tartaric acid and if $qq$, $l$-tartaric acid.

If the two halves of the molecule, each with one asymmetric carbon, were not identical, we would have two other optically active forms, $pq$ and $qp$. In the case of tartaric acid, however, the two halves *are* identical in structure, so that $pq$ and $qp$ are identical and, in each case, the optical activity of one half balances the optical activity of the other. The net result is optical *in*activity, and we have *meso*-tartaric acid.

It is not easy to see all this without careful structural formulas, which I will not plague you with. The crucial point to remember is that from 1874 right down to the present day, all questions of optical activity, no matter how involved, have been satisfactorily explained by a careful consideration of the tetrahedral carbon atom together with similar structures for other atoms. Although our knowledge of atomic structure has enormously expanded and grown vastly more subtle in the century since, Van't Hoff's geometrical picture remains as useful as ever.

Van't Hoff's paper dealing with the tetrahedral atom appeared in a Dutch journal in September 1874. Two months later, a somewhat similar paper appeared in a French journal. The author was a French chemist, Joseph Achille Le Bel, who was twenty-seven at the time.

The two young men worked it out independently, so that both are given equal credit and one usually speaks of the Van't Hoff-Le Bel theory.

The tetrahedral atom did not at once meet with the approval of all chemists. After all, there was still no direct evidence that atoms existed at all (and nothing direct enough

# THE 3-D MOLECULE

to be convincing was to come for another generation). To some of the older and more conservative chemists, therefore, the new view, placing atom bonds just so, smacked of mysticism.

In 1877, the German chemist Hermann Kolbe, then fifty-nine years old and full of renown, published a strong criticism of Van't Hoff and his views. It was quite within Kolbe's right to criticize, for it *could* be argued that the new view went beyond the foundations of chemistry as they then existed.

In fact, an essential part of the practical working of the scientific method is that new ideas be subjected to searching criticism. They must be jumped at and hammered down in fair and sporting fashion, for one of the tests of the value of the new idea is its ability to survive hard knocks.

Kolbe, however, was neither fair nor sporting. He characterized Van't Hoff as a 'practically unknown chemist', which had nothing to do with the case. Even more unforgivably, he sneered at him for holding a position at the Veterinary School of Utrecht, managing to refer to it three times in a short space, thus exhibiting a rather unlovely professorial snobbery.

Nevertheless, to those who think that the scientific 'establishment' has the power to quash useful advances permanently at the simple behest of conservatism and snobbery, let it be stated that the tetrahedral atom was adopted with reasonable speed. It worked so well that not all of Kolbe's sour fulminations could stop it and Van't Hoff's career went on untouched. (In fact, Van't Hoff rapidly became one of the leading physical chemists in the world and in 1901, when the Nobel prizes were established, the first award in chemistry went to him.)

Kolbe is today best known, perhaps, not for his own very real contributions to chemistry, but for his diatribe against Van't Hoff – which is reprinted to amuse the audience.\*

\* I was recently challenged to give my views on a book of far-out theory by someone who said he wanted my views *especially* if unfavorable, as he was making a collection which would someday, in

And again a new advance meant new problems. Once the structure of the carbon atom and its bonds had been worked out, and the details of molecules described in 3-D, a curious asymmetry turned out to exist in living tissue. That will be the subject of the next chapter.

---

hindsight, make very amusing reading. The book of far-out theory seemed like nonsense to me but I was aware of Kolbe's misfortune and I hesitated. But then I decided that I was not going to duck the issue out of fear for posterity's views. I thought the theories were worthless and I said so. However, I was polite about it. That much costs nothing.

## 5 – THE ASYMMETRY OF LIFE

Only yesterday (as I write this) I was on a Dayton, Ohio, talk show, by telephone, one of those talk shows where the listeners are encouraged to call in questions.

A young lady called in and said, 'Dr. Asimov, who, in your opinion, did the most to improve modern science fiction?'

I answered, after the barest hesitation, 'John W. Campbell, Jr.'*

Whereupon she said, 'Good! I'm Leslyn, his daughter.'

I carried on, of course, but inside I had a momentary dizzy spell. The reason for my second's hesitation in answering was that I had had to make a quick choice between two alternatives. I could have answered honestly and said, 'Campbell!' as I did; or I could have played it for laughs, as I so often do, and said, 'Me!' If I had had a visible audience and could have relied on hearing the laugh, I would undoubtedly have opted for the joke. As it was, with no possibility of a tangible reaction, I played it, thank goodness, straight – and avoided what would have been a terrible embarrassment.

Well, it sometimes happens, in science, that a person has a choice of two alternatives and has to face the possibility that his choice, whichever it is, will stamp itself indelibly on the field. If he guesses wrong, that wrongness may be impossible

* John Campbell, who died on July 11, 1971, was, in my opinion (and that of many others) the outstanding personality of all time in the field of science fiction. I owe a personal debt to him past all calculation. I have said this elsewhere. I wish to say so here.

to remove and will be a source of endless posthumous embarrassment.

Thus Benjamin Franklin once decided that there were two types of electric fluid and that one of them was mobile and one stationary. Thus some substances, when rubbed, gained an excess (+) of the mobile fluid, while others lost some of the mobile fluid and suffered a deficit (−). The one with the deficit showed the effect of the excess of the other, stationary fluid, so we could say that the two substances, (+) and (−), would show opposite electrical effects.

And so they do. An amber rod and a glass rod show opposite electrical effects when rubbed. (They attract each other, once charged, instead of repelling each other as like charges – two glass rods, for instance – would.) The question was: Which had the excess of the movable fluid and which the deficit; which was (+) and which was (−)?

There was absolutely no way of telling and Franklin was forced to guess. He guessed the amber had the excess, assigned it (+) and the glass he assigned (−). That set the standard. All other charges were traced back to Franklin's decision on amber *vs.* glass and to the present day it is usually assumed in electrical engineering that the current flows from the positive terminal to the negative.

By Franklin's standard the first two fundamental subatomic particles of ordinary matter were assigned their charge, too. The electron which tends to move toward the positive terminal is assigned (−); and the proton which is attracted to the electron is (+). They represent, in a sense, Franklin's two electric fluids, but, as it happens, it is the electron that is mobile and the proton that is relatively stationary, so that the current really flows from the negative terminal to the positive.

Franklin had had a fifty-fifty chance of guessing right, and he muffed it. Too bad. Fortunately, the wrong guess had no effect on the practical development of electrical technology or even on theory – but it always represents an irritating bit of non-neatness to neat-nuts like myself.

In this chapter, however, we will, in passing, mention

another fifty-fifty choice of alternatives and see how that worked out.

Once again, we are dealing with optical isomerism, the subject of the previous two chapters. Van't Hoff and Le Bel had shown (as I explained in Chapter 4) that if the four bonds of a carbon atom were attached to four different kinds of atoms or groups of atoms, that carbon atom was 'asymmetric'. The four attached groups could be attached in either of two possible configurations which were essentially different, one being the mirror image of the other.

A compound containing an asymmetric carbon atom can, in others words, be 'left-handed' or 'right-handed'.

As we might expect, nature has no left-right bias in this respect. Two compounds which differ, structurally, only in being left-handed or right-handed have identical chemical and physical properties and, when faced with conditions which are not themselves asymmetric, always react in the same way.

We might make an analogy to the right and left hand (or foot, or eye, or nostril, or upper canine). In each case the two organs have identical features and functions. What one can do the other can do and generally in equal fashion. The mirror imagery is not perfect, perhaps. The right and left hand of a given individual don't have mirror-image fingerprints, for instance. Also, most people use one hand with greater ease than the other – but that is because the brain itself is not perfectly symmetrical.

Chemical compounds, which are less complicated than the human hand, demonstrate left-right symmetry to a much higher degree of perfection than hands do. What a left-handed molecule can do, its right-handed brother can also do, and just as well.

(Of course, an equal mixture of right-handed and left-handed twins may have some properties which differ from those of either separately, as in the case of racemic acid and tartaric acid described in the previous chapter, but that's a different matter. A right hand and left hand clasped together can be easily distinguished from two rights – or two lefts –

clasped together, and because of the differing position of the thumbs, undoubtedly function differently.)

To see the significance of right-left symmetry, suppose you begin with a molecule that contains no asymmetric carbon and subject it to a chemical change that produces one. Thus, if a carbon has attached to it *abcc*, and you change one of the attached *c*'s to a *d*, so that the whole becomes *abcd*, a symmetric carbon becomes an asymmetric one.

The *d* can replace either of the two *c*'s. If it replaces one, there results a left-handed molecule and if it replaces the other, there results a right-handed molecule. The chances are exactly even; neither result is favored over the other.

Consequently, in any reaction of this sort, almost exactly equal numbers of each twin are produced. Any deviation from exact equality (and some deviation is to be expected in any chance process) would not be large enough to be detectable.

No matter what chemists do, short of introducing some asymmetric factor to begin with, they end up with symmetry. There seems no way of forcing Nature to make a right-left choice on the molecular level.

You can work the other way round. You can have a mixture containing equal numbers of the left-handed and right-handed mirror-image molecules, and subject that mixture to some physical or chemical effect (that is not, itself, asymmetric) which will alter the molecules. The altered molecules are such that they can be easily separated from the original. If the effect, whatever it is, destroys the left-hand molecule a little more rapidly or easily than the right-hand molecule (or vice versa), what will be left after a time, will show an excess of one or the other. The mixture will end by being at least slightly asymmetrical.

But that never happens either. You can't form molecular asymmetry out of a situation that is symmetrical to begin with.

I have been careful to rule out asymmetric effects till now, but suppose we decide to use one—

Suppose you have a substance made of two mirror-image twins in equal numbers; call them *b* and *d*, to use mirror-image letters. Next suppose you have another compound, which does *not* contain an asymmetric carbon atom, so that its molecules are symmetric. Call it *o*, a symmetric letter. If *o* combines with *b* and *d* to form an addition compound, then *bo* and *od* will be formed. These are still mirror images and can't be separated.

On the other hand, what happens if you have another compound which contains one or more asymmetric carbon atoms, so that it exists in right- and left-handed forms, and you actually have one or the other variety *only*? Call this *p*.

Again you form an addition compound and end up with *bp* and *pd*, which are *not* mirror images. (The mirror image of *bp* is *qd*, not *pd*.) The addition compounds, not being mirror images, have different properties and can be easily separated. Once the addition compounds are separated, each is broken down to *b* and *p*, or to *p* and *d*. The *p* is easily gotten rid of and the chemist is left with *b* and *d* in separate test tubes. He has two compounds, each of which is asymmetric and optically active, and this is called an 'asymmetric synthesis'.

You might very well ask, though, where a chemist gets the asymmetric *p* in the first place? If he can end with an asymmetric compound only when he begins with one, isn't he working in a circle? Where does the *first* asymmetric compound come from?

As it happens, it is easy to find compounds that are already asymmetric – but with an important restriction. He can find them only in connection with life. In fact, asymmetric compounds exist in nature *only* in living tissue or in matter that was once part of living tissue.

In fact, we can go farther than that. There are numerous molecules that have one or more asymmetric carbon atoms and that are to be found in living tissue. In every case only one of the optically active pairs is to be found there. If a left-hand compound is found in living tissue, the right-hand mirror image is *not*; if a right-hand compound is found in living tissue, the left-hand image is *not*.*

What's more, the choice between one twin and the other does not vary from species to species. If the left-handed twin is favored in the living tissue of any one species, it is favored in *all* living tissue of *all* species. All of earthly life makes use of only a single one of any molecule capable of existing as mirror-image twins, and always the same single one.

(This accounts, by the way, for the fact that Pasteur could separate the mirror-image components of racemic acid mechanically, as described in the previous chapter. Pasteur, being alive, was himself asymmetric.)

Is there perhaps some regularity to be found in which mirror-image twins will occur in tissue. At first glance, it doesn't seem so. Some compounds in living tissue are dextrorotatory and some are levorotatory and there seems no regularity to the matter. For instance, consider two very common sugars in living tissue: 'glucose' and 'fructose'. Both are made up of the same number of the same atoms and are very similar in properties. However, glucose is dextrorotatory and fructose, levorotatory, so that we have *d*-glucose and *l*-fructose.

Nor are these mirror images, I hasten to say. Each does have a mirror image, *l*-glucose and *d*-fructose, respectively, which do not occur in living tissue.

Once the Van't Hoff-Le Bel theory was advanced in 1874, something more than mere optical rotation was possible as a way of characterizing the mirror-image twins. Why not determine the actual configuration of the various groups about the asymmetric carbon atom and see if any regularity among the compounds found in living tissue follows from that?

This project was undertaken by the German chemist Emil Fischer, who began working with sugar molecules in the

* Actually, the non-occurring mirror images occasionally *do* occur, in specialized places and in very limited amounts. Their very trifling presences merely emphasize the general rule.

1880's. A molecule such as that of glucose has six carbon atoms, of which no less than four are asymmetric. Each one of the four can exist as a pair of mirror images, so that there are altogether sixteen different glucoselike compounds, arranged in eight pairs of mirror images.

To simplify matters, Fischer began with the simplest possible sugarlike compound, glyceraldehyde. It has three carbon atoms, of which only one is asymmetric. Glyceraldehyde therefore exists as just one pair of mirror-image twins, *d*-glyceraldehyde and *l*-glyceraldehyde.

The four different groups about the single asymmetric carbon atom in glyceraldehyde could be arranged in two different ways. Which arrangement should be assigned to the *d*-twin and which to the *l*-twin? Fischer had no way of telling, so he guessed! He assigned one arrangement, quite arbitrarily, to the *d*-glyceraldehyde and the other to the *l*-glyceraldehyde, establishing this standard in a paper he published in 1891.

(It wasn't till exactly sixty years later, in 1951, that it became possible to investigate molecules with sufficient subtlety to tell what the arrangement really was. This was accomplished by a team of Dutch investigators under J. M. Bijvoet, and they discovered that Fischer's fifty-fifty guess, unlike Franklin's, was *correct*.)

Fischer didn't stop there, of course. He began to build up, very carefully, more complicated sugar molecules, noting in every case what the arrangement must be. In every case, he could conclusively demonstrate that the structural arrangement of a complicated sugar with more than one asymmetric carbon atom was related to either the *d*-glyceraldehyde or the *l*-glyceraldehyde standard. Provided the atomic arrangements in the standard compounds were as he guessed they might be, he could work out the arrangements of all the others. (If he guessed wrong, then he would have to switch the arrangement in every sugar molecule to its mirror image – but, as eventually turned out, he hadn't guessed wrong.)

He found that although *d*-glyceraldehyde was dextrorotatory, some of the compounds related to it, structurally, were levorotatory. One could not predict from the structure alone the direction of optical rotation. Since lower-case letters had been used for direction of optical rotation, capital letters were used to indicate relationship. When a capital letter was used, the direction of rotation was indicated by (+) or (−), the former for dextro-, the latter for levo-.

Thus, since the glucose found in living tissue is related to $D$-glyceraldehyde and is dextrorotatory, it is called $D$-(+)-glucose. The fructose found in living tissue is also related to $D$-glyceraldehyde and is levorotatory, so it is $D$-(−)-fructose.

Here is something interesting. All the sugars found in living tissue, whether they turn the plane of polarized light in one direction or the other, are related to $D$-glyceraldehyde. They are all members of the '$D$-series'. To put it more dramatically, the sugars of life are all right-handed.*

But why?

If we seek the reason for any regularity in the structure of compounds in living tissue, we are bound to look at enzymes. All the compounds synthesized in living tissue are synthesized through the mediation of enzyme molecules, and all enzyme molecules are asymmetric.

We must ask, then, as to the nature of the asymmetry of enzymes.

All enzyme molecules are proteins. Protein molecules are made up of chains of amino acids which come in some twenty varieties. All twenty varieties are closely related in structure. In each case there is a central carbon atom to which are attached: (1) a hydrogen atom, (2) an amino group, (3) a carboxyl group, (4) any one of twenty different groups which may be lumped together as 'side chains'.

In the case of the simplest of the amino acids, 'glycine', the side chain is another hydrogen atom, so that the central carbon atom is attached to only three different groups. For that reason, glycine is not asymmetric and is not optically active.

* Minor exceptions? A substance related to $L$-(−)-glucose is found in streptomycin.

In the case of all the other amino acids, the side chain reppresents a fourth different group attached to the central carbon atom, which means that the central carbon is asymmetric and that each amino acid, except glycine, can exist in two forms, one the mirror image of the other. And, in fact, each amino acid exists in living tissue in only one of the two forms; and the same form is found, in each case, in all living tissue of any kind.

But which form? Some amino acids in the naturally occurring form are dextrorotatory and some are levorotatory, but you can't go by that. Instead, you must work out their structural nature with reference to the glyceraldehyde standard.

When this is done, it turns out that, *without exception*, all naturally occurring amino acids in all living tissue of whatever kind are of the *L*-series.*

We can therefore eliminate all questions as to why this form of some sugar (or other compound) exists in tissue and not its mirror image, and zero in on the amino acids. From them, everything else follows, so we can ask: Why are all the amino acids of the *L*-series?

It isn't hard to answer why all the amino acids belong to the same series. When amino acids hook together to form a protein molecule, the side chains stick out on this side or that and some of them are very bulky. The protein molecules do not have room to spare for them.

If the amino acid chain were to consist of both *L*-amino acids and *D*-amino acids, there would be frequent occasions when an *L*-amino acid would be immediately followed by a *D*-amino acid. In that case, the side chains would stick out on the same side and would, in many cases, seriously interfere with each other. If, on the other hand, the chain consisted of *L*-amino acids only, the side chains would stick out first to one side, then to the other, alternately. There would then be more room available and a protein molecule could more easily form.

* Well, almost. There are some *D*-series amino acids found in very specialized locations, in the cell walls of certain bacteria, for instance.

*But* the same thing would be true if the chain consisted of *D*-amino acids only. In fact, there is no reason to think that proteins consisting of *D*-amino acids only would be in any way different in form or function from those that now exist, that organisms made up of such *D*-proteins would be in any way inferior to those that now exist, that a whole ecology based on *D*-organisms would be in any way less viable than the system which does exist on Earth.

The question, therefore, arises: Why one rather than the other? Why has Earth developed an *L*-ecology, rather than a *D*-ecology?

The simplest possible explanation (and therefore the one which is perhaps most likely to be true) is through the working of sheer randomness.

In the lifeless primordial ocean, individual more complex molecules were steadily being built up out of less complex precursors thanks to energy sources such as the ultraviolet radiation of the Sun. Among these molecules being built up were *L*-amino acids and *D*-amino acids.* These come together to form chains, such chains being built up most easily out of all one form or all the other, so that both *D*-chains and *L*-chains would exist.

Eventually, some chains would be complex enough to have enzymatic properties and could co-operate, perhaps, with nucleic acids that would also be forming. (Nucleic acids contain five-carbon sugars in their molecules, which are *always* of the *D*-series.) It may be that, through sheer circumstance, an *L*-amino acid chain was first to reach the necessary complexity and, in combination with nucleic acid, began multiplying. (It is characteristic of life that it is based on molecules capable of forming replicas of themselves.)

In that way, the proto-life molecule, using itself as a model, could form many times more *L*-amino acid chains than could be formed by chance alone. The *L*-ecology would

* Since 1951, chemists have been trying to duplicate primordial conditions and have formed amino acids in this fashion – but always the *D*- and *L*-forms in equal quantities.

have got the first foothold and, being self-perpetuating, would never let go. The decision between $L$ and $D$ would thus be made at the very beginning of the history of life.

It might just as well have gone the other way, too, so that if we were to study many Earthlike life-bearing planets, we might find that about half of them bore a $D$-ecology and half an $L$-ecology.

Since food from $D$-organisms could be digested and assimilated only with difficulty, if at all, by $L$-organisms such as ourselves, and since it might set up serious, or even fatal, allergic manifestations, human exploration of the Galaxy might then face a particular danger. A planet might be a very paradise but if its life forms tested out $D$ it would be unsuitable for colonization.

But need we rely on pure randomness? There are some non-life sources of asymmetry. There is a kind of polarized light, called 'circularly polarized light', which can be viewed as either a left-handed screw or a right-handed screw.

A particular variety of such light, being asymmetric, would affect one mirror-image compound more than its twin. A chemist beginning with an equal mixture of the two mirror images would end with one slightly in excess. He would go from symmetry to asymmetry without the intervention of life. Usually, though, he ends with only some 0·5 per cent of the amount of asymmetry he would get if he had one of the images only.

Still, one can imagine a source of circularly polarized light on the primordial Earth, say through the reflection of sunlight from the ocean surface. The light might be harder on the $D$-amino acids than on the $L$-amino acids. The $D$-amino acids would be harder to form and easier to break down once formed. In that case there would be a kind of built-in bias in favor of the $L$-ecology.

The catch is, though, that there seems no reason why the circularly polarized light should be formed left-handed rather than right-handed. If it is formed in both ways equally, as is to be expected, there will be no bias.

But something new has turned up.

A Hungarian botanist named Garay (I don't have his first name) reported in 1968 that an amino acid solution bombarded with energetic electrons from strontium-90, did not decompose equally. The *D*-form decomposed perceptibly more quickly than the *L*-form.

Why?

One possibility is this. When the beta particles are slowed down by passage through the solution, they emit circularly polarized gamma rays. If the gamma rays were produced in equal amounts of left-handed and right-handed forms this wouldn't matter, but are they?

As I explained in Chapter 2, the law of parity breaks down in weak interactions, and it is these which involve the electron. The breakdown means that the electron is *not* symmetric with respect to right and left. It is left-handed, so to speak. Consequently, the gamma rays it produces are left-circularly polarized and that means *D*-amino acids are less easily formed and more easily destroyed once formed.

It would follow, then, that because of the non-conservation of parity there is an ingrained bias as far as optical isomers are concerned. In any Galaxy (or universe) made up of matter, in which electrons and protons dominate, we may expect a certain preponderance of *L*-ecologies among the life-containing planets.

On the other hand, in any Galaxy (or universe) made up of antimatter, in which positrons and antiprotons dominate, we may expect preponderance of *D*-ecologies among the life-containing planets.

Of course, this postulated connection between non-conservation of parity and the asymmetry of life is, as yet, highly tentative, but I am emotionally drawn to it. I firmly believe that everything in the universe is interconnected, that knowledge is one; and it seems dramatically right to me to have a discovery concerning the non-conservation of the law of parity, which seems so ivory-towerish and far-removed, serve to explain something so fundamental about life, about man, about you and me.

## B – The Problem of Oceans

# 6 – THE THALASSOGENS

Cocktail parties bring out the worst in me in the way of self-righteousness, for I don't drink.

This isn't a question of morality, you understand. It's just that I don't particularly like the taste of liquor and that even small quantities induce blotches and shortness of breath. Anyway, without ever touching a drop, I can be as hilariously drunk as anyone in the room – and no hangovers afterward.

The only trouble is that people won't let it go at that. They stand around and hound me. 'Are you *sure* you won't have something?' they ask for the fifteenth time.

What's more, when I do get thirsty, I have to go over to the bartender, make sure no one is listening, and then ask in a stage whisper if I can have some water.

First, I have to convince him that I really want water. Then I have to persuade him that I want a large glass without ice. I generally fail. Not listening, he picks out a cocktail glass and hands me water-on-the-rocks which means I have about five cubic centimeters of fluid and must then stand there, moodily, swirling ice cubes and wishing they would melt.

It's no wonder I get nasty. The other evening at a cocktail party one of those present was inveighing against marijuana. 'Ninety-two per cent of heroin users,' he said, 'began with pot.'

I was on his side, actually, for I am against the use of drugs, but I eyed the glass of liquor he was holding and said, 'Are you a social drinker?'

'Of course,' he said.

'Well,' said I, 'every single alcoholic who ever existed began as a social drinker.'

## THE THALASSOGENS

Anyway, there's nothing wrong with water. It's a great beverage and a very unusual substance in addition.

For instance, the six most common elements in the universe as a whole are thought to be hydrogen, helium, oxygen, neon, nitrogen, and carbon, in that order. Out of every ten thousand atoms in the universe about 9,200 are hydrogen, 790 are helium, 5 are oxygen, 2 are neon, 2 are nitrogen, and 1 is carbon. All the rest make up an insignificant scattering and for many purposes can be simply ignored.

With this information on hand, we can ask ourselves what the most common compound (*i.e.*, a substance with a molecule made up of two or more different kinds of atoms) in the universe is. It stands to reason that the most common compound would be one with a small, very stable molecule made up of atoms of the two most common elements.

Since helium atoms don't form parts of any molecules at all, that leaves hydrogen and oxygen as the most common compound-forming elements in the universe. One atom of each can combine to form 'hydroxyl' (OH), which has been detected in the interstellar spaces of our galaxy and of at least one other. It can only exist in rarefied media such as that in space. Two hydrogen atoms and one oxygen form water ($H_2O$), and that can exist at planetary densities – and is undoubtedly the most common such compound in the universe.

Naturally, water wouldn't be common everywhere. It wouldn't exist at all in any normal star, of course. The molecule breaks up at stellar temperatures. On too-small planetary bodies, water molecules would be too light and flitting to be held by the feeble gravitational force. Some might be held by chemical forces to the rocky crust, but this would represent a very small percentage of the total potential. It is not surprising then that the Moon, Mars, and, undoubtedly, Mercury are relatively dry.

On giant planets such as Jupiter and Saturn, where the gravitational field is intense and the temperature is low, there is a much more representative sampling of the material of the universe and surely water is by far the most common compound on such worlds.

Earth stands in an intermediate position. It is small enough and warm enough to have lost most of the water it might have possessed at the start. More likely, it failed to gather most of it in the first place out of the swirling cloud of dust and gas from which the planet formed. Even so, water on Earth is extremely plentiful.

In fact, in two respects, Earth's water is absolutely unique. In the first place, water is *by far* the most common liquid on Earth. Indeed, it is the only liquid on Earth present in quantity. (What is in second place? Petroleum, perhaps.)

Secondly, water is the only substance on Earth present, in quantity, in all three phases, solid, liquid, and gas. Not only is there an ocean full of water, but there are polar caps of miles-deep ice, and there is water vapor making up a major (if variable) part of the atmosphere.

The question, Gentle Readers, is this, then: Can any substance other than water serve? Can a planet exist with a large ocean of any substance other than water?

To answer that question, let's consider the requirements:

(1) The ocean substance must be a plentiful component of the universe mixture. We can imagine oceans of liquid mercury, or liquid fluorine or liquid carbon tetrachloride, but we can't realistically imagine any planet with these particular substances present in such quantities as to spread out into oceans.

(2) The ocean substance must have a prominent liquid phase. For instance, the Martian polar caps may well be frozen carbon dioxide, but there is no liquid carbon dioxide phase at Martian atmospheric pressure. The solid carbon dioxide vaporizes directly to gas, so there would be no carbon dioxide ocean even if there were enough carbon dioxide to form one.

(3) Ideally, we would want a substance whose liquid phase could be transformed with reasonable ease to either solid or gas, if we are to make possible those properties of Earth's ocean which lead to ice caps, clouds, rain and snow. Thus an

ocean of liquid gallium at the temperatures of water's boiling point, for instance, might produce gallium 'ice caps' with ease, but at that temperature, gallium's vapor pressure would be so low that there would be no gallium vapor in the air to speak of, no gallium clouds, no gallium rain. On the other hand, if we had an ocean of liquid helium at a temperature of 2° above absolute zero (*i.e.*, 2°K.) there would be plenty of helium vapor in the atmosphere (indeed, that would make up almost all the atmosphere) and helium rain would be common, but there is likely to be no helium ice or snow because solid helium doesn't form, even at absolute zero, except under considerable pressure, and we would be hard put to design a planet with sufficient atmospheric pressure at 2°K. to do the job.

In considering the requirements, let's begin with the first – presence in oceanic quantities. For that, we had better work with the top six elements only: hydrogen, helium, oxygen, neon, nitrogen, and carbon. Any substance made up of anything but these six elements (singly or in combination) might have many virtues but would simply not be present in sufficiently overwhelming a quantity to make up an ocean composed entirely or nearly entirely of itself.*

Of these six elements, two, helium and neon, can exist in elemental form only. A third, hydrogen, can form compounds, but exists in such overwhelming quantities that on any planet capable of collecting more than a trace of it (*i.e.*, on Jupiter, as opposed to Earth) it must exist mostly in elemental form for sheer lack of sufficient quantities of other elements with which to combine.

As for oxygen, nitrogen, and carbon, these, in the presence of a vast preponderance of hydrogen will exist only in combination with as much hydrogen as possible. Oxgen will exist as water ($H_2O$); nitrogen, as ammonia ($H_3N$); and carbon, as methane ($H_4C$).

---

* There is one conceivable exception on an Earthlike planet. Silicon dioxide is present in oceanic quantities but it is a solid and wouldn't be a liquid under anything but white heat. Scratch silicon dioxide.

This gives us our list of the six possible thalassogens:* hydrogen, helium, water, neon, ammonia, and methane, in order of decreasing quantity.

The next step is to consider each in connection with its liquid phase. At ordinary pressures, equivalent to that produced by Earth's atmosphere, each has a clear-cut boiling-point temperature, above which it exists only as a gas. This boiling point can be increased when pressure is increased, but let's ignore that complication, and consider the boiling point, in degrees above absolute zero, at ordinary pressure.

It turns out that the boiling points of helium, hydrogen, and neon are, respectively, 4·2°K., 20·3°K., and 27·3°K.

But keep in mind that even distant Pluto has a surface temperature estimated to be roughly 60°K. In fact, I wonder if any sizable planet, such as the outer members of our solar system, can ever have extremely low temperatures. Internal heat arising from radioactivity must be sufficient to keep the surface temperature at Plutonian levels, at least, even in the complete absence of any sun. (Jupiter, for instance, according to a recent report I've seen, radiates three to four times as much heat as it receives from the Sun.)

In short, then, for any reasonable planet we can design, the temperature is going to be too high for the presence, in quantity, of helium, hydrogen, or neon in the liquid phase. Scratch them from the list and we have only three thalassogens left: methane, ammonia, and water.

And what are their boiling points? Why, respectively, 111·7°K., 239·8, and 373·2.

If we consider these three, we come to these conclusions:

(1) Water is the most common and is therefore the most likely to form an ocean.

(2) Since methane is liquid across a range of 23 degrees, ammonia across 44 degrees, and water across 100 degrees, water, of the three, has by far the broadest temperature

---

* This is a word I have just made up. It is from Greek words ('sea-producers') and I define it as 'a substance capable of forming a planetary ocean'.

range for the liquid phase and, in its ocean-forming propensities, is least sensitive to temperature deviation.

(3) Most important of all, water forms its oceans at a higher temperature than the other two. You might expect methane oceans on a planet like Neptune or ammonia oceans on a planet like Jupiter. Only water, however, *only* water, could possibly form an ocean on an inner planet like Earth.

Well then, we depend for the existence of our ocean, and therefore for the existence of life, on the fact that water happens to have its liquid range at a far higher temperature than that of any other possible thalassogen. Is that just the way the ball bounces or is there something interesting to be wrung out of the water molecule?

Let's see—

When atoms combine to form molecules, the bond between them is formed through a kind of tug-of-war over the outermost electrons in those atoms. In many cases, one type of atom has the capacity to hold on to one or two electrons over and above those it normally possesses. Given half a chance it will grab on to such electrons. Since the atom itself is electrically neutral (positive charges in the interior, balancing negative charges on the outskirts) and since every electron has a negative charge, an atom which is capable of taking on one or more additional electrons then carries a net negative charge. Elements made up of atoms capable of doing this are therefore characterized as 'electronegative'.

The most electronegative of the elements, by far, is fluorine. Following it, in order, are oxygen, nitrogen, chlorine, and bromine. These are the only strong electronegative elements.

Some atoms on the other hand have no strong ability to latch on to additional electrons. Indeed, they find it difficult to hold on to the electrons they normally possess and have a considerable tendency to give up one or two. Given half a chance they will do so. Once they lose such negatively charged electrons, what remains of the atom has a net positive charge. Such atoms are therefore 'electropositive'.

Most of the elements tend to be somewhat electropositive. The most electropositive elements are the alkali metals, of which sodium and potassium are the most common representatives. Calcium, magnesium, aluminum, and zinc are other examples of strongly electropositive elements.

When an electropositive element, like sodium, meets an electronegative one like chlorine, the sodium atom freely gives up an electron, which the chlorine atom as freely takes. What is left is a sodium atom with a positive charge (a sodium ion) and a chlorine atom with a negative charge (a chloride ion). The attraction between the two ions is the strong pull of an electromagnetic force and this is called 'electrovalence'. A number of chloride ions cluster around each sodium ion and a number of sodium ions cluster around each chloride ion. The result is an intricate and very orderly array of ions that hang on to each other tightly.

The commonest way of pulling ions apart is to use heat. All ions, no matter how firmly held in place by some sort of attraction, are vibrating about that place. This vibration is related to temperature. The higher the temperature, the more energetic the vibration. If the temperature is high enough, the vibration becomes violent enough to pull the ions apart, however strong the electromagnetic force between them, and the substance then melts. (In the liquid phase, the ions are no longer held firmly in place and they move about freely.)

Nevertheless the temperature, by ordinary standards, must be quite high before the strong attractions between the sodium ions and the chloride ions can be overcome. Sodium chloride (ordinary table salt) has a comparatively high melting point, therefore – 1074°K. (For orientation, a pleasant spring day with the temperature at 70°F. is at 294°K.)

Still higher temperatures are required to pull the ions apart altogether and send them in pairs (one sodium ion and one chloride ion) into the nearly total independence of the gas phase, so that the boiling point of sodium chloride is 1686°K.

This is more or less true of all electrovalent compounds,

which form by the transfer of one or more electrons from one atom to another. Molybdenum oxide has a melting point of 2893°K. and a boiling point of 5070°K.

What happens, though, when one electropositive element meets another? Sodium atoms, for instance, can form bonds among themselves by allowing the outermost electron each possesses (and which they hold on to only very loosely) to be shared among them all. This is a stabler situation than would exist if each were responsible for its own outermost electron only, as in sodium gas. Consequently sodium atoms cling together and sodium is a solid at ordinary temperatures. To be sure, it doesn't take much to pull the atoms apart and sodium melts at a temperature of 370°K., just under that of boiling water. It doesn't boil, though, and obtain complete atomic independence till 1153°K.

(Those outermost electrons wander easily from atom to atom. Their existence accounts for the fact that sodium, and metals generally, conduct heat and electricity so much better than non-metals.)

Metals made up of less electropositive atoms get together more snugly and some of them end up by forming bonds as tight as those of any electrovalent compound. Tungsten metal has a melting point of 3640°K. and a boiling point of 6150°K.

Yet though metallic atoms fit together well, there is a greater tendency for them to transfer electrons to the electronegative atoms, particularly to oxygen, which is by far the most common of all the strongly electronegative elements. For this reason, there is virtually no free metal in the Earth's crust.*

In general, then, we can say that metals and electrovalent compounds are so high-melting as to offer no chance of a liquid phase at any reasonable planetary temperature, up to and including that of Mercury. Those few which might (like

---

* Earth has a metallic core because it contains so much iron that there just aren't enough electronegative atoms to take care of it all. The metallic excess, denser than the oxygen-containing electrovalent compounds, settled to the Earth's center in the soft, youthful days of the planet.

sodium metal or tin tetrachloride) cannot possibly be present in large enough quantity to form an ocean.

So we must look for something else. What happens if one electronegative atom meets another? What happens if one fluorine atom meets another, for instance? Each of the fluorine atoms can handle one electron over and above its usual assignment, but neither is in a position to give up one of its own in order to satisfy the other. What does happen is that each atom allows the other a share in one of its own electrons. There is a two-electron pool to which each contributes and in which each shares. Both fluorine atoms are then satisfied.

In order for this pool to exist, though, the two fluorine atoms must remain at close quarters! To pull them apart takes a lot of effort, for it means breaking up that two-electron pool. Consequently, under ordinary circumstances, fluorine in elementary form exists in molecules made up of atom pairs ($F_2$). The temperature must rise well over 1300°K. even to begin to break up the fluorine molecule and shake the individual atoms apart. The attraction between atoms represented by shared electrons is called a 'covalent bond'.

Two fluorine atoms, once they have formed their two-electron pool, have no reason to share any electrons with any other atoms, much less transfer electrons to them or even receive electrons from them. The two-electron pool completely satisfies their electron needs. Consequently, when one fluorine molecule meets another fluorine molecule, they bounce off each other, with very little tendency to stick together.

If there were no tendency to stick together at all, the fluorine molecule would remain independent of its neighbors however far down the temperature might drop. The molecules would move more and more sluggishly, bounce off one another more and more feebly, but they would never stick.

However, there are what are called 'Van der Waals forces', named for the Dutch chemist who first studied them. Without going into the matter in detail we can simply say that there are weak attractive forces between atoms or molecules even when there is no outright electron transfer or electron sharing.

Thanks to Van der Waals forces, fluorine molecules are slightly sticky, and if the temperature drops low enough, the energy that keeps them moving will not be great enough to make them break away after colliding. Fluorine will condense to a liquid.

The boiling point of liquid fluorine is 85°K. If the temperature drops further still, the fluorine molecules lock firmly into an orderly array and fluorine becomes a solid. The melting point of solid fluorine is 50°K.

The same thing happens with the other electronegative elements. Chlorine, oxygen, and nitrogen also form electron pools between two atoms. We therefore have chlorine molecules, oxygen molecules, and nitrogen molecules, each made up of atom pairs ($Cl_2$, $O_2$, and $N_2$). Even hydrogen atoms, which are not particularly electronegative, form molecules by pairs ($H_2$).

In every case the melting and boiling points are low, with the exact value depending on the strength of the Van der Waals forces. Hydrogen, with its very small atoms, possesses a liquid range at a considerably lower temperature than that of fluorine. The boiling point of liquid hydrogen is 21°K. and the freezing point of solid hydrogen is 14°K.

A few varieties of atom happen to possess a satisfactory number of electrons to begin with. They have little tendency to give up any electrons they have and still less to accept additional electrons from outside. They do not therefore tend to form compounds. These are the so-called 'noble gases'.

There are six of these altogether and, of them, the three with the largest atoms can form compounds (not very stable ones) with the most electronegative elements, such as fluorine and oxygen. The three with the smallest atoms – argon, neon, and helium (in order of decreasing size) – won't do even that much under any conditions yet discovered. Nor will they form electron pools among themselves. They remain in sullen isolation as individual atoms.

Yet they, too, experience the mutual attraction of Van der Waals forces and, if cooled sufficiently, become liquids. The smaller the atom, the smaller the forces and the more strongly cooled they must be to liquefy.

Helium, with the smallest atoms of the noble gases, experiences such small attractions that of all known substances it is the most difficult to liquefy. The boiling point of liquid helium is phenomenally low, only 4·2°K. Solid helium doesn't exist at all, even at 0°K. (absolute zero), except under considerable pressure.

So far, though, these gaseous substances I have discussed, that are covalent in nature, and that have liquid ranges far down the temperature scale, are all elements – elements that either exist in the form of isolated atoms, as in the case of helium, or as isolated two-atom molecules, as in the case of hydrogen.

Is it possible for molecules of two different atoms to be covalent in nature and to be low-melting and low-boiling. —Yes, it is!

Consider carbon. The carbon atom is neither strongly electropositive nor strongly electronegative. It has a tendency to form two-electron pools with each of four other atoms. It could form those pools with four other carbon atoms, each of which can form pools with three others, each of which with still three others, and so on idefinitely. In the end, uncounted trillions of carbon atoms may be sticking firmly together by way of strong covalent bonds. The result is that carbon has a higher melting point than that of any other known substance – nearly 4000°K.

But the carbon atom may form a two-electron pool with each of four different hydrogen atoms. The hydrogen atoms can only form one two-electron pool apiece and so that ends it. The entire molecule consists of a carbon atom surrounded by four hydrogen atoms ($H_4C$), and this is methane.

Methane molecules have little attraction for each other except by way of weak Van der Waals forces. The boiling point of liquid methane is 112°K., and the melting point of solid methane is 89°K.

THE THALASSOGENS 89

Similarly, a carbon atom can form a molecule with one oxygen atom. This would be carbon monoxide (CO). Its boiling point and melting point are, respectively, 83°K. and 67°K.

Now we can come to a general conclusion. Unlike metallic substances and electrovalent compounds, covalent compounds have low melting points and boiling points and only they can conceivably be thalassogens at reasonable planetary temperatures.

This gives us our first answer as to why water is a thalassogen at all: it is a covalent compound essentially. All right, that's something to begin with. Yet so many covalent compounds are, if anything, liquid at two low a range for planetary purposes and certainly for earthly purposes specifically. Why is liquid water so warm then?

One possibility rests in the fact that, in general, the larger the covalent atom or molecule, the stronger the Van der Waals forces and the higher the boiling point. Consider the following table, in which the size of the molecule is measured by its molecular weight (or, in the case of helium and neon, atomic weight).

| Substance | Atomic or Molecular Weight | Boiling Point (°K.) |
| --- | --- | --- |
| Hydrogen ($H_2$) | 2 | 17 |
| Helium (He) | 4 | 4 |
| Neon (Ne) | 20 | 27 |
| Nitrogen ($N_2$) | 28 | 77 |
| Carbon monoxide (CO) | 28 | 83 |
| Oxygen ($O_2$) | 32 | 90 |
| Fluorine ($F_2$) | 38 | 85 |
| Oxygen fluoride ($OF_2$) | 54 | 138 |
| Nitrogen fluoride ($NF_3$) | 71 | 153 |
| Chlorine ($Cl_2$) | 71 | 239 |
| Pentane ($C_5H_{12}$) | 72 | 309 |
| Chlorine heptoxide ($Cl_2O_7$) | 183 | 355 |

The table isn't perfect, for helium, which has a larger atomic weight than hydrogen's molecular weight, nevertheless has a lower boiling point than hydrogen. Then, too, fluorine, which has a larger molecule than oxygen has, is nevertheless lower-boiling. Still, the table seems to show that there is a kind of rough and ready relationship between molecular weight and boiling point in the case of covalent compounds.

We might conclude therefore that water, which has a boiling point at 373°K., ought to have a molecular weight somewhat higher, or at least not particularly lower, than chlorine heptoxide. Its molecular weight ought to be, say, 180, as a minimum.

Except that it isn't. The molecular weight of water is 18, just one tenth what it 'ought' to be.

Something, obviously, is terribly wrong – or right, perhaps, for it is to whatever causes this anomaly that we owe our life-giving ocean. What that wrongness/rightness might be we'll discuss in the next chapter.

# 7 – HOT WATER

One of the occupational hazards of popularizing the scientific view of the universe for the general public is the occasional collision with readers who prefer some variety of religious view of the universe instead. To reduce some wonderful phenomenon from the provenance of God to the blind consequence of some physical or chemical 'law' offends them, and their response, very often, is to accuse the science writer of atheism.

Thus, only yesterday, I received a letter from a lady which began by addressing me austerely as 'Dear Sir,' and then continued, somewhat less austerely, 'According to the Scriptures, and using the language of the Scriptures, you are a "fool".'

That aggrieved me, naturally, since while I am every bit as foolish (on occasion) as the next fellow, I hate to be told so. Besides, the accusation went beyond that of mere folly. It was obvious that the lady was referring to a certain well-known biblical quotation.

Among the hundred and fifty poems in the Book of Psalms, there are two, the fourteenth and the fifty-third, that are virtually identical, and in each case the first verse begins, 'The fool hath said in his heart, There is no God.'

What could I do? I decided that a scriptural reference deserves a scriptural reference, so I sent the nice lady a short note which said:

' "... whosoever shall say, Thou fool, shall be in danger of hell fire." Matthew 5: 22.'*

But, alas, having taken care of one correspondent, I must

* This is from the Sermon on the Mount, in case you don't recognize it.

now run the risk of offending others of those whom Robert Burns would refer to as the 'unco guid'. For, you see, water has amazing properties that seem to be perfectly designed for life. It would be so pious to look upon it as the workings of a benevolent and ingenious Maker, creating a Universe for the good of undeserving Man, and so prosaic to bring it down to the uncaring properties of atoms.

—Yet I will have to do the latter, since I am committed to the scientific view of the universe (pointing out to the reverent that they can easily suppose those uncaring properties to have been created by God).

In the previous chapter, I pointed out that water was the only possible thalassogen for a planet at the temperature of the Earth, the only compound that could possibly exist in sufficient quantity in the liquid phase to form an ocean.

To be liquid at the relatively low temperatures of Earth (as I explained), a substance would have to consist of covalent molecules, that is, molecules in which pairs of neighboring atoms more or less share electrons in neighborly fashion, rather than carrying through a transfer of one or more electrons from one atom to another, bodily.

In general, the larger the molecular weight of a covalent compound the higher the temperature range of its liquid phase. From that standpoint one might expect a substance which is liquid at water temperatures to have a molecular weight of perhaps 180. The molecular weight of water, however, is 18, just one tenth of what it 'ought' to be. On the basis of molecular weight, liquid water is surprisingly warm; it is 'hot water' indeed.

But why is that? Are we perhaps oversimplifying if we relate liquid-phase temperatures to molecular weight alone?

Well, at the end of the previous chapter I listed molecular weights and boiling points without any attempt to pick and choose among them. That is probably unfair, for substances are made up of different elements, and these differ greatly among themselves in chemical and physical properties. Yet elements exist in families, and within these, the members are quite similar. It might be best to stick to members of a particular family and see what regularities we can find there.

For instance, consider the six elements of the noble gas family, their atomic weights, and their boiling points.

Table 1

| Element | Atomic Weight | Boiling Point (°K.)* |
|---|---|---|
| Helium (He) | 4·0 | 4·2 |
| Neon (Ne) | 20·2 | 27·2 |
| Argon (Ar) | 39·9 | 87·4 |
| Krypton (Kr) | 83·8 | 120·2 |
| Xenon (Xe) | 131·3 | 166·0 |
| Radon (Rn) | 222·0 | 211·3 |

Here we have a smooth rise in boiling point with the atomic weight, which is what we would expect, looking at the matter in an unsophisticated way. After all, if the atoms grow heavier, it takes more energy in the form of heat to lift them away from each other and send them off separately in vapor form.

What if we shift to the four elements of another family, the halogens, a family as well defined as that of the noble gases (see Table 2). Here, too, the boiling point rises smoothly with

Table 2

| Element | Atomic Weight | Boiling Point (°K.) |
|---|---|---|
| Fluorine (F) | 19·0 | 85·0 |
| Chlorine (Cl) | 35·5 | 238·5 |
| Bromine (Br) | 79·9 | 331·9 |
| Iodine (I) | 126·9 | 457·5 |

the atomic weight. There is a fifth halogen, the last in the series, which is named astatine. It is a radioactive element and even its most long-lived nuclear variety (with an atomic weight of 210) has a half-life of but 8·3 hours. It has not yet

* '°K.' represents the 'absolute scale of temperature with the zero point ("absolute zero") at −273·16°C.'.

been obtained in quantities large enough to allow a clear boiling point determination, but I am willing to bet, sight unseen, and any reasonable sum, that its boiling point is somewhere in the neighborhood of 570°K.

While the progression is smooth within an element family, observe what happens when we cross the line. Compare Tables 1 and 2. Neon and fluorine are not very different in atomic weight but fluorine's boiling point is three times as high as neon's. This goes all the way down, each halogen having a boiling point approaching three times that of the noble gas with similar atomic weight.

Is it that atomic weight alone isn't the sole deciding factor? Of course not. There are other properties that play a role. The noble gas atoms are chemically inert and do not combine among themselves. They remain as separate atoms. The halogen atoms, on the other hand, because of their characteristic electron arrangements (different from those of the noble gas atoms) do combine in pairs. Fluorine is not composed of individual atoms, as neon is, but of molecules made up of two atoms apiece. Fluorine is $F_2$ and its molecular weight is 38·0. To consider the energy required to separate the fundamental particles of fluorine liquid into vapor, one ought to consider the weight of the molecule, not that of the atom. The molecular weight of fluorine is about that of the atomic weight of argon and, sure enough, the boiling point of fluorine is about that of argon.

If we could stop there, we'd be able to work up a hard-and-fast relationship between particle size (whether atomic or molecular) and boiling point. In science, though, it isn't fair to stop at any point where you find yourself with the answer you want. You have to be sporting enough to look further and try to spoil your own hypothesis.

That's not hard to do. Chlorine atoms combine by twos also, and chlorine is $Cl_2$, with a molecular weight of 71. That is distinctly less than the atomic weight of krypton, and yet the boiling point of chlorine is just twice that of krypton.

So we had better not try to cross the family lines in working up our theories. For the rest of the article I will stick to families and it will only be anomalies within the families that will receive our attention.

But let's see, is it only the boiling points that vary smoothly with atomic (or molecular) weight? Is the variation always direct, so that the measure grows larger as the weight goes up? Let's consider a third well-marked element family, that of the 'alkali metals' and take their melting points this time.

*Table 3*

| Element | Atomic Weight | Melting Point (°K.) |
|---|---|---|
| Lithium (Li) | 6·9 | 452 |
| Sodium (Na) | 23·0 | 371 |
| Potassium (K) | 39·1 | 337 |
| Rubidium (Rb) | 85·5 | 312 |
| Cesium (Cs) | 132·9 | 301 |

Cesium's melting point is down to 301°K., or 28·5°C., which means that it will melt on a hot summer day. There is also a sixth alkali metal, francium, which is radioactive, with its most long-lived nuclear variety (atomic weight, 223) having a half-life of only 21 minutes. Its melting point has not been determined but you can bet, though, it is very likely about 290°K. and that it will melt on a balmy spring day.

Other properties of other kinds vary in this regular fashion with atomic weights within element families, with values sometimes moving steadily upward and sometimes steadily downward.* The next question is, though, will this same sort of happy effect work within families of compounds – that is, substances with molecules made up of more than one kind of atom?

Consider molecules made up of carbon and hydrogen. These come in many varieties, because carbon atoms can link together in chains and rings. Suppose, then, we consider a single carbon atom combined with hydrogen, a chain

* It is only fair to say that a rigidly steady variation is not always found. There are exceptions. However, modern chemists can usually account for them, and we will have an example later in this article.

of two carbon atoms combined with hydrogen, a chain of three carbon atoms, four, and so on. The longer the chain the larger the molecular weight, and we can consider such a series of progressively larger molecules of very much the same kind to make up a family. What happens to the boiling point, in that case?

Table 4

| Compound | Molecular Weight | Boiling Point (°K.) |
| --- | --- | --- |
| Methane ($CH_4$) | 16·0 | 111·7 |
| Ethane ($C_2H_6$) | 30·1 | 184·5 |
| Propane ($C_3H_8$) | 44·1 | 228·7 |
| Butane ($C_4H_{10}$) | 58·1 | 273·7 |
| Pentane ($C_5H_{12}$) | 72·2 | 309 |
| Hexane ($C_6H_{14}$) | 86·2 | 341 |

As you see, boiling point rises smoothly with molecular weight in this case.

To be sure, the family of 'hydrocarbons' considered in Table 4 is one in which all the members have molecules made up of the same elements. Would it be possible to set up families in which at least one of the elements changes from member to member?

Thus, carbon is the first member of an element family of which the next three, in order of increasing atomic weight, are silicon (Si), germanium (Ge), and tin (Sn). An atom of each of these higher members can combine with four hydrogen atoms to form well-known compounds (silane, germane, and stannane, respectively) analogous to methane. Table 5 shows what happens to the boiling points there, and you see we get regularity in such a family, too.

The problem, then, of finding out why water has the high liquid-range temperatures it has, may become easier to handle if we work within some family of compounds that includes it.

Water molecules are made up of hydrogen and oxygen atoms ($H_2O$). Of these two elements, hydrogen is a loner and is not part of any clearly defined family (though it has cer-

## HOT WATER

### Table 5

| Compound | Molecular Weight | Boiling Point (°K.) |
|---|---|---|
| Methane (CH$_4$) | 16·0 | 111·7 |
| Silane (SiH$_4$) | 32·1 | 161·4 |
| Germane (GeH$_4$) | 76·6 | 184·7 |
| Stannane (SnH$_4$) | 122·7 | 221 |

tain relationship both to the halogens and to the alkali metals). Oxygen, on the other hand, is the first member of a family that includes sulfur (S), selenium (Se), and tellurium (Te) as later members. An atom of each of these three can combine with two hydrogen atoms to form molecules (H$_2$S, H$_2$Se, and H$_2$Te, respectively) that are analogous in structure to water molecules.

### Table 6

| Compound | Molecular Weight | Boiling Point (°K.) |
|---|---|---|
| Water (H$_2$O) | 18·0 | 373·2 |
| Hydrogen sulfide (H$_2$S) | 34·1 | 213·5 |
| Hydrogen selenide (H$_2$Se) | 81·0 | 231·7 |
| Hydrogen telluride (H$_2$Te) | 129·6 | 271·0 |

If we look at the last three members alone, we see that the boiling point goes up with molecular weight. *But water doesn't fit!* Its boiling point should be, judging from the rest, something like 200°K. or −73°C. Only the coldest polar days should suffice to liquefy its vapor and yet here it is, boiling something like 170 degrees higher than it should. Hot water, indeed.

There are two other compounds that, like water, don't fit their families in this respect.

A hydrogen atom will combine with one atom of any of the halogens. We can get hydrogen fluoride (HF), hydrogen chloride (HCl), hydrogen bromide (HBr), and hydrogen iodide (HI). The boiling points of the last three on the absolute scale are 188·2, 206·5, and 237·8, respectively. We might expect HF to have a boiling point about 170, but it doesn't. Its boiling point is 292·6 or about 120 degrees 'too high'.

Then, too, three hydrogen atoms will combine with one atom of the member of a family of elements that includes nitrogen (N), phosphorus (P), arsenic (As), and antimony (Sb). The compounds phosphine ($H_3P$), arsine ($H_3As$), and stibine ($H_3Sb$) have boiling points of 185·5, 218, and 256. On that basis, the first member of the series, ammonia ($H_3N$), ought to have a boiling point of about 150, but it doesn't. Its boiling point is 239·8, which is about ninety degrees 'too high'.

What, then, do these three too-high-boiling compounds, water ($H_2O$), ammonia ($H_3N$), and hydrogen fluoride (HF), have in common?

(1) All three are made up of molecules consisting of hydrogen atoms and one other kind of atom.

(2) The other atoms involved – oxygen, nitrogen, and fluorine – just happen to be the three most electronegative atoms there are; that is, the atoms most capable of snatching electrons from other atoms.

A fluorine atom, the most electronegative of all, can, for instance, take an electron away from a sodium atom altogether, assuming sole ownership and leaving the sodium atom utterly minus one electron.

The hydrogen atom is not quite such an easy mark. It holds on to its single electron more tightly than the sodium atom does to its one outermost electron. The fluorine atom does not take hydrogen's electron away altogether, but it does take over the lion's share of it. The electron, so to speak, is closer to the center of the fluorine atom than to the center of the hydrogen atom.

This means that if you imagine a line drawn down the center of the hydrogen fluoride molecule, with the hydrogen atom on one side and the fluorine atom on the other; the fluorine side, having more than its equal share of electrons, has what amounts to a small negative electric charge, while the hydrogen side has an equally small positive electric charge.

Much the same can be said of the water molecule and the

ammonia molecule. In each case, the side of the hydrogen atoms carries a small positive charge, while the side of the oxygen (or nitrogen) atom carries a small negative charge.

All three molecules are 'polar molecules'. That is, they have poles at which electric charge is concentrated.

This is not true of $H_2S$, for instance, which is otherwise so similar to $H_2O$ in structure. Sulfur just isn't as electronegative as oxygen and it cannot hog more than its fair share of the electrons of the hydrogen atoms. Hydrogen sulfide is therefore not particularly polar. Neither is hydrogen chloride or phosphine.

If we now consider polar molecules, those with a positively charged end and a negatively charged end, we must inevitably start thinking of the possibility of attraction between molecules. What if the positively charged end of one molecule should be near the negatively charged end of another molecule of the same kind? Would they not stick together a bit?

Yes, they would, particularly since the positively charged end involves the hydrogen atom. Why? Because the hydrogen atom is the smallest of all the atoms and its center can therefore be most closely approached. The strength of attraction between two oppositely charged objects varies inversely as the distance between them. The closer they come together, the stronger the attraction.

It follows, then, that the water molecule, the hydrogen fluoride molecule, and the ammonia molecule are 'sticky molecules'. They tend to line up positive end to negative end, and it takes significantly higher temperatures to pry them apart than if they were non-polar; that is, lacking the concentration of charge on two opposite sides, and held together only by the Van der Waals forces mentioned in the previous chapter.*

* Van der Waals forces are also the result of electrical asymmetry in atoms and molecules, with momentary concentrations of electric charge in one place or another. In non-polar molecules, however, the concentration-shifts from place to place lead to overall polarizations that are tiny indeed, much less than those in polar molecules, where the concentration of charge is persistent and definitely localized.

Usually, the water molecules are pictured with a hydrogen atom attached to the oxygen atom of its own molecule by a solid bond representing an ordinary chemical linkage, while it is attached to the oxygen atom of a neighboring molecule by a longer, dashed bond to indicate the electromagnetic attraction of opposite charges.

Because the hydrogen atom is thus between two oxygen atoms, one of its own and one of a neighboring molecule (or, in similar fashion, between two fluorine atoms, between two nitrogen atoms, between a nitrogen atom and an oxygen atom, and so on), the situation is commonly referred to as a 'hydrogen bond'.

The hydrogen bond is only about one twentieth as strong as an ordinary chemical bond, but that is enough to add up to 170 degrees to the temperature required to tear the molecules apart and set the liquid to boiling. Water molecules are sticky enough, thanks to the hydrogen bonds, to boil at 373°K. instead of 200°K., and that, combined with the fact that hydrogen and oxygen are the two most common compound-forming atoms in the universe, makes it possible for oceans of liquid to exist on a planet the temperature of the Earth.

What's more, it is because of the stickiness of the water molecules that it is possible for water to absorb so much heat for each degree rise in temperature or give off so much heat for each degree fall. We say, therefore, that water has an unusually high 'heat capacity'.

There is, similarly, an unusually high heat absorption at the melting point or boiling point, due to the necessity for breaking all those hydrogen bonds. That is, it takes much more heat than one would expect to convert ice at 273°K. to water at the same temperature, or to convert water at 373°K. to steam at the same temperature. Working in reverse, an unusual amount of heat is given off when steam condenses to water or water freezes to ice. (Water has an unusually high 'latent heat of fusion and vaporization', in other words.)

This is more than a mere matter of statistics. Water acts

as a huge heat-sponge. It takes up and gives off more heat than any other common substance for a given change of temperature, so that the ocean rises in temperature much more slowly under the beating rays of the Sun than the land does, and drops in temperature much more slowly in the absence of the Sun.

With a vast ocean of water on its surface, the Earth therefore has a much more equable temperature than it would without it. In the summer, the sluggishly warming ocean acts as a cooling device; in the winter, the sluggishly cooling ocean is a warming device. And if you want to see what that means in a practical sense, consider the temperature ranges over the day-night interval and the summer-winter interval of a land area far from any ameliorating stretch of ocean (North Dakota) with those of one that is surrounded by ocean on all sides (Ireland).

Since at any temperature, the evaporation of water absorbs more heat per gram of vapor formed than is true of any other common liquid, water is a particularly cheap and effective air-conditioning device.

Perspiration is almost pure water and as it evaporates a great deal of heat must be absorbed from the object closest to that water – which happens to be the skin on which the perspiration rests. In this way, the body is cooled.

Then, too, there is the matter of solvent properties. In a substance like sodium chloride (common salt), the sodium atoms lose an electron each to the chlorine atoms, which therefore gain an electron each. The sodium atoms carry a unit positive charge and the chlorine atoms a unit negative charge, and are hence called ions. The two sets of ions cling together through the attraction of opposite charges.*

When particles of salt are dumped into water, the presence of positive and negative poles on the water molecules sets up an electromagnetic field which tends to neutralize that which is set up by the charged sodium and chloride ions. The ions cling to each other with far less verse in the presence of

* The sodium chloride combination is much more polar than the water molecule is and this is reflected in its extremely high boiling point.

water than in the open air and have a pronounced tendency to fall apart and go swimming in the water on their own. To put it briefly, sodium chloride dissolves in water.

So do a surprising variety of other electrovalent compounds, that is, compounds made up of oppositely charged ions after the fashion of sodium chloride.

Polar compounds, which are not built up of outright ions but have molecules with separated concentrations of charge (like water itself), also lose a considerable part of their tendency to cling together in the presence of water and therefore tend to dissolve. This includes many common substances of importance to life which have the oxygen-hydrogen or nitrogen-hydrogen linkage that makes polarization possible.

This includes various alcohols, sugars, amines, and other organic compounds.

No other liquid is so versatile a solvent as water; no other liquid can dissolve appreciable quantities of so wide a variety of substances. To be sure, though, water cannot dissolve appreciable amounts of all electrovalent compounds, since electrovalency is not the only property that is important. And, of course, it cannot dissolve non-polar compounds such as hydrocarbons, fats, sterols, and so on.

The importance of water's versatile solvent action is this—

The body's most important substances, the proteins and the nucleic acids, together with its most important fuels, the starches and sugars, are loaded with oxygen-hydrogen and nitrogen-hydrogen linkages and, if not polar altogether, have important polar regions within their molecules. Such compounds can therefore dissolve in water or, at least, can attach water molecules intimately to various portions of their structure and undergo changes in connection with these attached water molecules.

In short, the body's chemistry can go on against the intimacy of a water background. This background is so essential to life as we know it, that life could only have reasonably begun in the ocean, and now, even where it has adapted itself to dry land, the tissues remain approximately 70 per cent water.

So consider water. Consider its high liquid-range temperatures, its capacity to act as a temperature-ameliorating heat-sponge and as an efficient air conditioner, its ability to dissolve a wide variety of substances and, therefore, to act as a medium within which the reactions necessary to life can proceed, and you may well say, 'Surely, this is no accident. Surely water is a substance that has been carefully designed to meet the needs of life.'

But that is placing the cart before the horse, I'm afraid. Water existed, to begin with, as a substance of certain properties, and life evolved to fit those properties. Had water had other properties, life would have evolved to fit those other properties. If water had a lower liquid-range temperature, for instance, life might have evolved on Jupiter. And if water had not existed at all, life might have evolved to fit some other substance altogether.

In every case, though, life would have evolved so neatly to fit whatever was at hand, that any form of that life high enough to consider the situation with sufficient subtlety would well feel justified in believing that intelligent and purposeful supernatural design was involved in something which, actually, the blind and random forces of evolution had produced.

And I suppose my delightful lady correspondent, if she were so hardy as to read through this essay carefully, would but feel herself further justified in her belief about the relationship of scriptural language to myself.

But what can I do? I call the situation as I see it.

# 8 – COLD WATER

About half a year ago (as I write this) I was hurrying through the wintriness of New York City. There was no snow on the ground but it was cold and I was hastening for haven. As I was crossing the street, my foot came down upon a manhole cover and a fraction of a second later I had made hard and full-length contact with the ground.

It was the hardest fall I had ever taken and my first thought, as I lay there, was one of regret, for it felt as though I had broken my left tibia and in all my thirty-plus years I had never before broken a bone. I ought to have lain there and waited for help, but I had to struggle to my feet for two reasons:

For one thing, I was hoping desperately the bone was not broken and if I could get to my feet it wouldn't be. Secondly, I wanted to find out why I had fallen, since I am usually reasonably sure-footed.

I found I could stand. My left leg was banged up below the knee but the bone was intact, even though my suit (my *best* suit) was not. I further found (more in anger than in sorrow) that the manhole cover was frosted over with a thin layer of slippery ice. What had laid me low was the fact that the ice was quite transparent and that without close inspection, the manhole cover seemed bare and safe.

I had to hobble onward, at that moment, toward my hotel room, which was four infinitely long blocks away, and there was no time to muse on what had happened and make an article out of it. By now, though, the bitterness of the time has been somewhat assuaged, and I am ready. So here, O Gentle Reader, is the result—

To the ancients, one of the remarkable things about ice,

perhaps the most remarkable, was the very property that had caused my near disaster – its transparency. To the Greeks, ice was *krystallos* from kryos, meaning 'frost', so the first strong impression seems to have been left by its manner of formation.

Once that was established, however, another property supervened and the word came to be more significant for the connotation of transparency than of cold. After all, anything at all could be cold, but in ancient times few objects were known that were at the same time solid but not opaque.

It followed, then, that when pieces of quartz were discovered, and found to be transparent, they were called *krystallos*, too, and were considered (at first) to be a form of ice that had been subjected to such intense cold as to have attained permanent solidity and an inability ever to melt again.

Then the word achieved still another change of connotation. One interesting fact about transparent quartz was its surprising regularity in shape. It had plane faces that met to form clearly defined angles and edges. Consequently, *krystallos* came to mean any solid with such a regular geometry. From this came our modern word 'crystal'.

Nevertheless, the older meaning of transparency persists in vestigial fashion. One still hears of the 'crystalline spheres' which held the planets in the old Ptolemaic cosmology. This was not because they consisted of solid crystals; heavens, no. It was because they were perfectly transparent so they could not be seen.

And in modern times, the fortuneteller, gazing mystically into a glass sphere, is pretending to see something in her 'crystal ball'. This is not because the sphere is crystalline in the modern sense, for glass happens to be one of the very few common solids that is *not* crystalline (and, therefore, not truly solid), but because it is transparent.

And yet, none of that really represents the true wonder of ice. It may seem to have wonders enough. Its mere existence as 'hard water' may seem amazing and paradoxical enough to the life-long inhabitants of tropic climes, and its coldness and transparency may be of interest, but all that is really nothing.

Consider instead something that is often remarked on, to the point, in fact, where it becomes something of a cliché. Have you never heard a statement such as this one: 'Like an iceberg, nine tenths of the significance of the remark was hidden'?

Like an iceberg!

Being a non-traveler, I have never seen a real iceberg, but if I were on a ship and one hove into view (at, I hope, a safe distance) I am sure that the passengers, crowding against the rail to see it, would say to one another, 'Just imagine, Mabel (or Harry), nine tenths of that iceberg is under water.'

Then I would say, 'That's not surprising, ladies and gentlemen. The surprising thing is that one tenth of that iceberg is above water.' Naturally, that would mean I would start getting those queer looks that would indicate once again (oh, how many times!) how much of a nut I appear to be to my beloved fellow-creatures.

But it's true—

In general the density of any substance increases as temperature goes down. The lower the temperature, the more slowly the atoms or molecules of a gas move, the less forcefully they bounce off each other, and the closer they can crowd to each other. When the kinetic energy of the gas molecules is insufficient to overcome the attractive forces between the molecules (see the previous two chapters), the gas liquefies.

In liquids, the molecules are in virtual contact, but they have enough energy to slip and slide past each other freely. They also vibrate and keep each other at greater distances than would be the case if all were absolutely motionless. As the temperature drops, the vibrations decrease in force and amplitude and the molecules settle a bit closer together. The density continues to increase.

Eventually, the energy of vibration isn't enough to keep the molecules slipping and sliding. They settle into a fixed position and the substance solidifies. The settling is more compact than is possible (usually) in the liquid form but there is still vibration about the fixed position. As the temperature continues to drop, the vibrations continue to die down until they are reduced to a minimum at the temperature of absolute zero ($-273 \cdot 1°C$.). It is then that density is at a maximum.

To summarize— As a general rule, there is an increase in density with decrease in temperature. There is a sudden sharp increase in density when a gas becomes a liquid* and another, but lesser, sharp increase when a liquid becomes a solid. This means that the solid form of a substance, being denser than the liquid form of that same substance, will not float in the liquid form.

As an example, liquid hydrogen has a density of about $0 \cdot 071$ grams per cubic centimeter, but solid hydrogen has a density of about $0 \cdot 086$ grams per cubic centimeter. If a cubic centimeter of solid hydrogen were completely immersed in liquid hydrogen it would still weigh $0 \cdot 015$ grams and would be pulled downward by gravity. Sinking slowly (against the resistance of the liquid hydrogen) but definitely, it would eventually reach the bottom of the container, or the bottom of the ocean, if there were that much liquid hydrogen.

(You might suspect that the solid hydrogen would melt on the way downward, but not if the ocean of liquid hydrogen were at its freezing point – and we'll suppose it is.)

In the same way, solid iron would sink downward through an ocean of liquid iron, solid mercury through liquid mercury, solid sodium chloride through liquid sodium chloride, and so on. This is so general a situation that if you took a thousand solids at random, you would be very likely to find that in each case the solid form would sink through the liquid form and you would be tempted to make that a universal rule.

—But you can't, for there are exceptions.

And of these, by far the most important one is water.

\* Except at the 'critical temperature', something which need not concern us now.

At 100°C. (water's boiling point under ordinary conditions), water is as un-dense as it can be and still remain liquid. Its density then is about 0·958 grams per cubic centimeter. As the temperature drops the density rises: 0·965 at 90°C., 0·985 at still lower temperature, and so on until at 4°C., it is 1·000 grams per cubic centimeter.

To put it another way, a single gram of water has a volume of 1·043 cubic centimeters at 100°C., but contracts to a volume of 1·000 cubic centimeters at 4°C.

Judging from what is true of other substances, we would have every right to expect that this increase in density and decrease in volume would continue as the temperature dropped below 4°C. It does *not*!

The temperature of 4°C.* represents a point of maximum density for liquid water. As the temperature drops below that, the density starts to decrease again (very slightly, to be sure) and by the time one reaches 0°C., the density is 0·9999 grams per cubic centimeter, so that a gram of water takes up 1·0001 cubic centimeters. The difference in density at 0°C. as compared with that at 4°C. is trifling, but it is in the 'wrong' direction, and that makes it crucial.

At 0°C. water freezes if further heat is withdrawn, and by everything we learn from other solidifications we would have a right to expect a sharp increase of density. We would be wrong! There is a sharp *decrease* in density.

Whereas water at 0°C. has, as I said, a density of 0·9999 grams per cubic centimeter, it freezes into ice at 0°C. with a density of only about 0·92 grams per cubic centimeter.

If a cubic centimeter of ice is completely immersed in water, with both at a temperature of 0°C., then the weight of the ice is −0·08 grams and there is, so to speak, a negative gravitational effect upon it. It therefore rises to the surface of the water. The rise continues till only enough of it is submerged to displace its own weight (as measured in air) of the denser, liquid water. Since a cubic centimeter of ice at 0°C. weighs 0·92 grams and it takes only 0·92 cubic centimeters of water at 0°C. to weigh 0·92 grams, it turns out that when the

* 3·98°C., to be more accurate.

ice is floating, 92 per cent of its substance is below water and 8 per cent is above.

What we would ordinarily expect, judging from almost all other solids immersed in their own liquid form, is that 100 per cent of the ice would be submerged and 0 per cent exposed. It follows, then, as I said earlier, that the surprising thing is not that so much of an iceberg is invisible, but that so much of it (or, indeed, any of it at all) is exposed.

Well, why is that?

Let's begin with ice. In ordinary ice, each water molecule has four other molecules surrounding it with great precision of orientation. The hydrogen atom of each water molecule is pointed in the direction of the oxygen atom of a neighbor and this orientation is maintained through the small electrostatic attraction involved in the hydrogen bond (as described in the previous chapter).

The hydrogen bond is weak and does not suffice to draw the molecules very close together. The molecules remain unusually far apart, therefore, and if a scale model is built of the molecular structure of ice, it is seen that there are enough spaces between the molecules to make up a very finely ordered array of 'holes'. Nothing visible, you understand, for the holes are only about an atom or so in diameter.

Still, this makes ice less dense that it would be if there were a closer array of molecules.

As the temperature of the ice rises, its molecules vibrate and move still farther apart, so that its density falls, reaching a minimum of the aforementioned 0·92 grams per cubic centimeter at 0°C. At that temperature of 0°C., however, the molecular vibration has reached the point where it just balances the attractive forces between the molecules. If further heat is added, the molecules can break free and can begin to slip and slide past each other. In doing so, however, some of them fall into the holes.

As ice melts, then, the tendency to decrease the density through increased vibrational energy is countered by the disappearance of the holes, and more than countered. For that reason, liquid water is 8 per cent denser at 0°C. than solid water is.

Even in water at 0°C., however, the loose molecular arrangement in ice hasn't utterly vanished. As the temperature rises still higher, there is still a slow disappearance of the last few lingering holes and it is not till a temperature of 4°C. is reached that so few of them are left that they can no longer exert a dominating effect on the density change. At temperatures higher than 4°C., the energy of molecular vibration increases and density decreases steadily as it 'ought' to do.

The importance of this density anomaly in water simply can't be exaggerated. Consider what happens to a moderately sized lake, for instance, during a cold winter.

The temperature of the water gradually drops from its mild warmth of the summer. Naturally, it is the water at the surface that cools first, becomes denser, and sinks, forcing up the warmer water at the bottom, so that it can, in turn, cool and sink. In this way the entire body of water cools, and would cool all the way to 0°C. if the density continued to increase steadily as temperature dropped.

As it is, though, when a temperature of 4°C. is reached, a further cooling of the surface water makes it slightly less dense! It does *not* sink, but floats on the warmer water below. The surface water drops in temperature all the way to 0°C., but heat leaves the lower depths only slowly and those depths remain somewhat warmer than 0°C.

It is the water at the suface, then, that experiences freezing, and the ice, being less dense than water, remains floating. If the cold weather continues long enough, the entire layer of surface water freezes and forms a solid coating of ice that may become thick and strong indeed (to the satisfaction of ice skaters).

But ice is a good insulator of heat, and the thicker it is, the more effective an insulator it is. As it thickens, the deeper layers of water (still liquid) lose heat through the ice to the air above more and more slowly; and more and more slowly does the ice layer thicken further. In short, in any winter that is likely to occur on Earth, a sizable lake will never freeze solidly all the way to the bottom. This means that life-forms in it can survive through the winter.

What's more, when the warm weather returns, it is the surface ice that receives the brunt of the Sun's heat. It melts and the liquid water beneath is at once exposed, so that the lake quickly becomes liquid throughout once more.

What would happen, though, if water were like other substances. In cooling, there would be a continual sinking of cooler water all the way down to 0°C., so that the entire body of the lake would be at that temperature eventually. It would have a tendency to freeze at every point, and any ice that formed near the surface of the lake would sink at once if there were still liquid below it. A winter that under present circumstances would only suffice to form a thick scum of ice on a lake would be enough to freeze that same lake solid, top to bottom, if water were like other substances.

Then, when warm weather came, the surface of the frozen lake would melt, but the water that formed would insulate the deeper layers of ice from the Sun's heat. The thicker the layer of liquid water, the more slowly the Sun's heat would penetrate to the ice below and the more slowly would the deeper ice melt. Through an ordinary summer such as we experience on Earth, a solidly frozen lake would never melt all through. Most of it would remain permanently frozen.

The same would hold true for rivers and for the polar oceans. Indeed, if water were suddenly to change its density characteristics, each winter would see further ice form and sink to the ocean abyss to remain permanently frozen thereafter. Eventually, all of Earth would be a mass of ice-bound land, with a thin layer of water on the surface of the tropic ocean.

Even though such an Earth would be at the distance from the Sun it now is, and would receive the amount of solar energy it now does, it would be a frigid world and life as we know it would not have formed. It follows then that life depends on the hydrogen bond not merely for the reasons I outlined in the last chapter, but because of the loose structure it gives ice.

There's another way to break down the holes in ice besides raising the temperature. Why not simply squeeze the ice together under pressure? To be sure, it takes enormous pressures to squeeze out the holes to the point where ice is as dense as water. (When water is allowed to fill a sealed container tightly and then made to freeze, it exerts an outward pressure equal to the pressure it would take to compress ice to the density of water – and the container breaks.)

Still, high pressures can be produced in the laboratory. About 1900, a German physicist, Gustav Tammann, began to make use of such high pressures, and beginning in 1912, an American physicist, Percy W. Bridgman, carried the matter much further.

In this way, it was found that there were many forms of ice.

In any solid there is an orderly arrangement of molecules and there is always the possibility of a variety of different arrangements under different conditions. Some arrangements are more compact than others and these would be favored by high pressures and low temperatures.

Thus, under ordinary temperatures and pressures, ordinary ice (which we can call Ice I) is the only variety that can exist. As the pressure is increased, however, two other forms are found, Ice II at temperatures below $-35°C.$, and Ice III at temperatures between $-35°C.$ and $-20°C.$

If the pressure is raised still further, Ice V is formed. (There is no Ice IV; it was reported but proved to be a case of mistaken observation and was dropped; but not before Ice V had been reported.)

If the pressure is raised still further, Ice VI and Ice VII are formed. Whereas all other forms of ice exist only at $0°C.$ and below, Ice VI and Ice VII can exist at temperatures above $0°C.$, though only at enormous pressures.

In fact, at a pressure of 20,000 kilograms per square centimeter (one and a half million times the pressure of the atmosphere), Ice VII will exist at temperatures above 100°C., the boiling point of liquid water under ordinary conditions.

All these high-pressure forms of ice are denser than liquid water, as you would expect, for the holes have been squeezed out of them. Indeed, of all known forms of ice, only Ice I, the ordinary variety, is less dense than liquid water.

It would follow that if any of the forms of ice other than Ice I could form in the oceans, they would sink to the bottom and gradually accumulate.

In one of his excellent novels, Kurt Vonnegut hypothesized a mythical 'Ice IX,' which could exist at the ocean bottoms and which would form spontaneously if only some small quantity existed as a 'seed.' The hero had such a small piece and, of course, it got into the ocean to bring about the final catastrophe.

Is there really a chance of that? No. Any form of ice but Ice I can only exist at enormous pressures. Even the least high-pressure ice (Ice II and Ice III) can exist only at pressures more than two thousand times that of the atmosphere. If such pressures could be attained at the bottom of the ocean (they can't), a further requirement would be that the temperature be well below $-20°C$. (It isn't.)

You can see, further, that no form of ice other than Ice I could exist in someone's pocket. If any other ice were formed and the high pressure required to form it were removed, the ice would instantly expand to Ice I with explosive violence.

That still leaves one thing to discuss. Though solid forms of a substance can (and often do) exist in a variety of crystalline forms, liquid and gaseous forms do not. In liquids and gases there is not, generally speaking, any orderly array of molecules, and one does not find varieties of disorder.

But in 1965, a Soviet scientist, B. V. Deryagin, studied liquid water in very thin capillary tubes and found some of it to possess most unusual properties. For one thing, its density was 1·4 times that to be expected of ordinary water. Its boiling point was extraordinarily high and it could be heated up to 500°C. before ceasing to be liquid. It could be cooled down to −40°C. before turning into a glassy solid.

The report was largely disbelieved in the West, where there is almost automatically skepticism toward any unusual finding that emerges outside the charmed circle of nations prominent in nineteenth-century science.

However, when Americans repeated Deryagin's work, they found, much to their own surprise, that they got the same results and could even see droplets of the anomalous form of liquid water – droplets so small they could be made out only under a microscope.

What was behind this?

Water molecules, while slipping and sliding around each other, do tend to take up the hydrogen bond orientation, as in ice. This happens over very small volumes and for very brief periods, but it is enough to make liquid water behave as though it consisted of submicroscopic particles of ice that form and unform with super-speed.

The 'ice' never forms over a volume large enough and for a time long enough to make the holes significant and cause water to be as un-dense as ice, but it does keep the water molecules far enough apart to allow hydrogen bonds to form and unform. Liquid water is therefore less dense than it might be.

Suppose, though, that pressure is placed on water in such a way that molecules are forced closer together while in the hydrogen bond orientation. With neighboring molecules unusually close, the hydrogen bond would be much stronger than ordinary and would, indeed, approach an ordinary chemical bond in strength. Molecule after molecule would fall into place and, thanks to the unusually strong hydrogen bond attractions, they would make up a kind of giant molecule built up out of the small water-molecule units.

When small units build up a giant molecule in this fashion, the small units are said to 'polymerize' and the giant molecule is a 'polymer.' The new form of water was therefore spoken of as 'polymerized water' or, for short, 'polywater.'

In polywater, the molecules are in orderly array, some-

## COLD WATER

thing as in ice, but in much more compact fashion, and certainly without the holes. Not only does this compact array of water molecules produce a substance considerably more dense than ice, but considerably more dense than ordinary liquid water as well.

What's more, because the molecules are held more tightly together, it takes a much higher temperature than 100°C. to tear them apart and make polywater boil. It also takes a much lower temperature than 0°C. to force the molecules apart into the less compact array of ordinary ice. Other unusual properties of polywater are also easily explained on the basis of the compact array of molecules.

Apparently, polywater does not form from under ordinary increases in pressure, but does form in the constricted volume of tiny capillary tubes. Biologists at once began to wonder whether within the constricted volume of tissue cells, polywater also formed; and whether some of the properties of life could not be most easily explained in terms of polywater.

I wish I could end the matter here, with this glamorous discovery and the still more glamorous speculation, but I can't. The trouble is that many chemists remain skeptical of the whole business.

It is possible, after all, that investigators have been misled by the chance of solution of the glass from the tubes in which the polywater was being studied. If it were not pure liquid water they were studying but tiny volumes of glass solution, all bets were off.

Indeed, one chemist recently prepared a solution of silicic acid (something which could form when water is in contact with glass) and reported it to possess the very properties of polywater.

So it may be that polywater is a false alarm, after all.

*C – The Problem of Numbers and Lines*

# 9 – PRIME QUALITY

Not long ago I got a letter from a young amateur mathematician which offered me a proof that the number of primes was infinite and asked, first, if the proof were valid, and, second, if it had ever been worked out before.

I answered that first, the proof was a valid and elegant one but second, that Euclid had worked out the same proof just about word for word, in 300 B.C.

Alas, alas, this is the fate of almost every single one of us amateur mathematicians almost every single time. Anything we work out that is true is not new; anything we work out that is new is not true. —And yet, if we work out what is true, from a standing start, without ever having had it worked out for us, than I maintain it to be a feat of note. It may not advance mathematics, but it is a triumph of the intellect just the same.

I told my young correspondent this and now I would like to tell you about the proof and about a few other things.

First, what is a prime or, more correctly, a 'prime number'? A prime is any number that cannot be expressed as the product of two numbers, each smaller than itself. Thus, since $15 = 3 \times 5$, 15 is *not* a prime. On the other hand, 13 cannot be expressed as a product of smaller numbers and *is* therefore a prime. Of course, $13 = 13 \times 1$, but 13 is not smaller than 13, so that this multiplication does not count. *Any* number can be expressed as itself multiplied by 1, whether it is prime or not ($15 = 15 \times 1$, for instance), and this sort of business is no distinction.

Another way of putting it is that a prime number cannot be divided evenly ('has no factors') other than by itself and by 1. Thus 15 can be divided evenly by either 3 or 5, in *addition* to being divisible by 15 and 1; but 13 can be divided *only* by 13 and 1. So again, 15 is not a prime and 13 is.

Well then, what numbers are prime? Alas, that is not an easy question to answer. There is no general way of telling a prime number just by looking at it.

There are certain rules for telling if a particular number is *not* a prime, but that is not the same thing. For instance, 287,444,409,786 is *not* a prime. I can tell that at a glance. What's more, 287,444,409,785 is *not* a prime, either, and I can tell that at a glance, too. But is 287,444,409,787 a prime? All I can tell at a glance is that it *may* be a prime; but also it may *not*. There is no way I can tell for certain unless I look it up in a table – assuming that I have a table that gives me all the prime numbers up to a trillion. If I don't have such a table, and I don't, I have to sit down with pen and paper and try to find a factor.

Is there any systematic way of finding all the primes up to some finite limit. Yes, indeed, there is. Write down all the numbers from 1 to 100. (I'd do it for you here, but I'd waste space, and it will be good exercise for you if *you* do it.)

The first number is 1 but that is *not* a prime by definition. The reason for that is that in multiplication – which is the way we have of distinguishing primes from non-primes – the number 1 has the unique property of not changing a product. Thus 15 could be written as $5 \times 3$, or as $5 \times 3 \times 1$, or, indeed, as $5 \times 3 \times 1 \times 1 \times 1 \times 1$ ... and so on forever. By simply agreeing to eliminate 1 from the list of primes, we eliminate the possibility of a tail of 1's, and get rid of some nasty complications in the theoretical work on primes. No other number acts like 1 in this respect and no other number requires special treatment.

We next come to 2, which is a prime since it has no factors other than itself and 1. Let's eliminate every number in our list that can be divided by 2 (and is therefore non-prime) and to do that we need only cross out every second number after 2. This means we cross out 4, 6, 8, 10, and so on, up to and including 100. You can check for yourself that these numbers are not prime, since $4 = 2 \times 2$; $6 = 2 \times 3$; $8 = 2 \times 4$, and so on.

We look at our list of numbers and find that the smallest number *not* crossed out is 3. This is a prime since 3 has no factors other than itself and 1. So we begin with 3 and cross out every third number after it: 6, 9, 12, 15, and so on, up to and including 99. Some of the numbers, 6 and 12, for instance, were already crossed out when we were dealing with 2, but that's all right; cross them out again. The numbers now crossed out are all divisible by 3 and are therefore *not* prime: $6 = 3 \times 2$; $9 = 3 \times 3$, and so on.

The next number not crossed out is 5, and you cross out every fifth number after it. Then 7, and you cross out every seventh number after it. Then 11, then 13, and so on. By the time you reach 47 and proceed to cross out the 47th number after it (94), you find you have crossed out every number below 100 that you can. The next available number is 53, but if you try to cross out the 53rd number after it, that is 106, which is beyond the end of the list.

So you have left the following numbers under 100 which are not crossed out: 2, 3, 5, 7, 11, 13, 17, 19, 23, 29, 31, 37, 41, 43, 47, 53, 59, 61, 67, 71, 73, 79, 83, 89, 97.

These are the twenty-five prime numbers below 100. If you memorize them you will be able to tell at a glance whether any particular number under 100 is a prime or not, just by knowing whether it is or is not on the list I just gave you.

Is there a simple connection among all these numbers, some formula that will give only the primes up to 100 and no other numbers? Even if you could work out such a formula, it wouldn't help you, for it would break down as we proceed above 100, for after all, we can, if we want to, continue to use the same system of stopping at each uncrossed number and counting off every one that is its own number after it. We would then find out that above 100, there are prime numbers such as 101, 103, 107, 109, 113, 127, and so on.

If we had written all the numbers up to 1,000,000,000,000,

we would eventually have worked out all the primes up to that point and we would have determined, mechanically and without flaw (provided we make no mistake in counting), whether the number I gave you previously, 287,444,409,787, is or is not a prime.

This perfect system for finding all the prime numbers up to any finite number, however large, is called 'the sieve of Eratosthenes,' because the Greek scholar Eratosthenes first used it somewhere about 230 B.C.*

There is one trouble with the sieve of Eratosthenes and that is that it takes an unconscionable length of time. Working it out through 100 isn't bad, but time yourself working it through 1,000 or through 10,000 and you'll agree that it soon piles up prohibitively.

But wait. After all, you keep piling up more and more prime numbers and each one sieves out some of all the still higher numbers remaining. This means that a larger and larger percentage of those still higher numbers is crossed out, doesn't it?

Yes, it does. There are twenty-five primes under 100, as I just pointed out, but only twenty-one primes between 100 and 200, and sixteen primes between 200 and 300. This dwindling is an irregular thing and sometimes the number jumps, but on the whole, the percentage of primes does dwindle – there are only eleven primes between 1,300 and 1,400.

Well, then, do the primes ever come to a complete halt?

Put it another way. As one goes up the line of numbers, there are longer and longer intervals, *on the average*, between primes. That is, there are longer and longer lists of successive non-primes. The longest successive stretch of non-primes under 30 is five: 24, 25, 26, 27, and 28. There are seven successive non-primes between 89 and 97; thirteen successive

* When Frederik Pohl (the well-known science fiction writer and editor) was young, he worked out the sieve of Eratosthenes all by himself and was most chagrined to find out he had been anticipated. But one needs no more evidence than that to demonstrate Fred's brightness. Working it out independently (simple though it seems after it is explained) was more than I was ever able to do.

non-primes between 113 and 127, and so on. If you go high enough, you will find a hundred successive non-primes, a thousand successive non-primes, ten thousand successive non-primes, and so on.

You can find (in theory) any number of non-primes in succession no matter how high a number you name, if you proceed along the list of numbers long enough. *But*, and this is a big 'but,' is there ever a time when the number of non-primes in succession is infinite? If so, then after a certain point in the list of numbers, *all* the remaining numbers will be non-prime. The number marking that 'certain point' would be the largest prime number possible.

What we are asking now, then, is whether the number of primes is infinite or whether there is, instead, some one prime that is the largest of all, with nothing prime beyond it.

Your first thought might be to work out the sieve of Eratosthenes till you reach a number beyond which you can see that everything higher is crossed out. That, however, is impossible. No matter how high you go, and how long a vista thereafter seems to be non-prime, you can never possibly tell whether there is or is not another prime somewhere (perhaps a trillion numbers further) up ahead.

No, you must use logical deduction instead.

Let's consider a non-prime number that is a product of prime numbers: say, $57 = 19 \times 3$. Now let's add 1 to 57 and make it 58. The number 58 is *not* divisible by 3, since if you try the division you get 19 with a remainder of 1; nor is it divisible by 19, for that will give you 3 with a remainder of 1. This is not to say that 58 is not divisible by any number at all, for it is divisible by 2 and by 29 ($58 = 2 \times 29$).

You can see, however, that any number that is the product of two or more smaller numbers, is no longer divisible by any of *those* numbers if its value is increased by 1. To put it in symbols:

If $N = P \times Q \times R \ldots$, then $N+1$ is *not* divisible by either P or Q or R or any other factor of N.

Well, then, suppose you begin with the smallest prime, 2, and consider the product of all the successive primes up to some point. Begin with the two smallest primes: $2 \times 3 = 6$. If you add 1 to the product you get 7, which cannot be divided by either 2 or 3. As a matter of fact, 7 is a prime number. You go next to $(2 \times 3 \times 5) + 1 = 31$ and that's a prime. Then $(2 \times 3 \times 5 \times 7) + 1 = 211$, and $(2 \times 3 \times 5 \times 7 \times 11) + 1 = 2311$, and both 211 and 2311 are primes.

If we then try $(2 \times 3 \times 5 \times 7 \times 11 \times 13) + 1$, we get 30,031. That, actually, is *not* a prime number. However, neither 2, 3, 5, 7, 11, nor 13 (which represent *all* the primes up to 13) are among its factors, so any primes that must be multiplied to make 30,031, must be higher than 13. And, indeed 30,031 $= 59 \times 509$.

We can say, as a general rule, that 1 plus the product of any number of successive primes, beginning with 2 and ending with P, is either a prime itself and is therefore certainly higher than P, or is a product of prime numbers all of which are higher than P. And since this is true for any value of P, there can be no highest prime, since a mechanism exists for finding a still higher one, no matter how high P is. And that, in turn, means that the number of primes is infinite.

This, in essence, is the proof Euclid presented, and it is the proof my young correspondent worked out independently.

The next problem is this: Granted that the number of primes is infinite, is there any formula that has as its solution all the primes and none of the non-primes, so that we can say: Any number that is a solution of this formula is a prime; all others are not? You see, to determine whether 287,444,409,787 is a prime or not by the sieve of Eratosthenes, which will surely tell you, you must work your way up through all the lower numbers. You can't skip. A 'prime-formula' will enable you to crank in 287,444,409,787 directly and tell you whether it is prime or not.

Alas, there is no such formula, and it is not likely that any can ever be found (although I am not sure that it has been *proven* that none can be found). The order of primes along the list of numbers is utterly irregular and no mathematician has ever been able to work out any order, however complicated, which would make a 'prime-formula,' however complicated, possible.

Let's lower our sights then. Is it possible to work up some useful formula that will give us not *every* prime, but at least *only* primes? We could then grind out an infinite series of known primes by turning a formula-crank, even though we know we are skipping an infinite quantity of other primes.

Again, no (except for some specialized non-useful cases). No matter how we try to find a useful system that will yield primes only, non-primes will *always* sneak in. For instance, you might think that adding 1 to the product of successive primes beginning with 2 might yield only prime numbers. The numbers I got this way a little earlier in the article were 7, 31, 211, 2311 – all primes! But then, the next in the series was 30,031, and that was *not* a prime.

Formulas have been worked out in which the value $n$ was substituted by the numbers 1, 2, 3, and so on, with prime values obtained for every value up to $n = 40$. And then for $n = 41$, a non-prime will pop out.

So let's lower our sights again. Is there any system that will allow us to crank out only non-primes? Non-primes may not be interesting but at at least we can get rid of them and study a group of remaining numbers that will be denser in primes.

Yes! At last we have something to which the answer is, yes! In working out the sieve of Eratosthenes, for instance, perhaps you noticed that in crossing out every second number after 2, you crossed out *only* numbers that ended with 2, 4, 6, 8, and 0, and that you crossed out *every* number that ended with 2, 4, 6, 8, 0. This means that any number, no matter how long and formidable, even if it has a trillion digits, is *not* a prime if the last digit is 2, 4, 6, 8, or 0; if it's an 'even number,' in other words.

Since exactly half of all the numbers in any finite successive list end in these digits, that means that all primes (except for 2 itself, of course) must exist in the other half – the odd numbers.

Then again, when you begin with 5 and cross out every

fifth number, you cross out *only* numbers that end with 5 and 0, and *every* number that ends with 5 and 0. Numbers ending with 0 are already taken care of, but now we can eliminate any number from the list of possible primes if the last digit is 5 (except for 5 itself, of course).

This means we need look for primes (other than 2 and 5) *only* in those numbers that end in the digits 1, 3, 7, or 9. This means that in any successive list of numbers we can eliminate 60 per cent and look for primes only in the remaining 40 per cent.

Of course, if we take into account not a finite successive list of numbers (say from 1 to 1,000,000,000,000) but *all* numbers, the 40 per cent that may contain primes is still infinite and still contains an infinite number of primes – and an infinite number of non-primes, too. Restricting the places in which we look for primes doesn't help us in the ultimate problem of finding all the primes by some mechanical method easier than the sieve of Eratosthenes but at least it clears away some of the underbrush.

Of course, there are other possible eliminations. Any number, no matter how long and complicated, whose digits add up to a sum divisible by 3 is itself divisible by 3 and is not a prime. However, digit adding is tedious, so let's restrict ourselves to just looking at the last digit. The trick of looking at the last digit is the only elimination device that is simple enough to be pleasing. Is there anything we can do to improve the situation that exists?

To answer that, let us ask what the magic is of 2 and 5 that enables them to make their mark on the final digit. The answer is easy. Our number system is based on 10 and $10 = 2 \times 5$. What we have to do is find a number that is the product of two separate primes that is *smaller* than 10. Maybe we can then crowd the 'magic' into a smaller area.

Only one number smaller than 10 will do and that is $6 = 2 \times 3$.

All numbers are either multiples of 6 or, on being divided by 6, leave remainders that are equal to 1, 2, 3, 4, or 5. There are no other possibilities. This means that any number is of the class $6n$, $6n+1$, $6n+2$, $6n+3$, $6n+4$, or $6n+5$. Of these, any number of the form $6n$ cannot be a prime since it is divisible by both 2 and 3 ($6n = 2 \times 3n = 3 \times 2n$). Any number of the form $6n+2$ or $6n+4$ is divisible by 2 and any number of the form $6n+3$ is divisible by 3.

That means that all primes (except 2 and 3) must be of the forms $6n+1$ or $6n+5$. Since $6n+5$ is equivalent to $6n-1$, we might say that all prime numbers are either one more or one less than a multiple of 6.

Suppose, then, we make a list of multiples of 6: 6, 12, 18, 24, 30, 36, 42, 48, 54, 60, 66, 72, 78, 84, 90, 96, 102 . . .

With that as a guide, we could next make a double list of all numbers one less than these multiples, and one more, with bold face for those numbers, which are prime:
**5, 11, 17, 23, 29,** 35, **41, 47, 53, 59,** 65, **71,** 77, **83, 89,** 95, **101** . . . **7, 13, 19,** 25, **31, 37, 43,** 49, 55, **61, 67, 73, 79,** 85, 91, **97, 103** . . .

As you see, the numbers in the list occur in pairs of which one is 2 more than the other (with a multiple of 6 in between). You might think, after looking at the list above, that at least one of each pair must be a prime and that that imposes some kind of additional order on the primes. That is not so, unfortunately. At least one of each pair is a prime as far as we've gone, but if you go further, you will find that in the pair 119, 121, neither one is a prime. The number 119, which is $6 \times 20 - 1$, is equal to $7 \times 17$ and 121, which is $6 \times 20 + 1$, is equal to $11 \times 11$. The higher up you go the more common the non-prime pairs get.

Sometimes only the upper and smaller number of the pair is a prime, as in **23** and 25; sometimes only the lower and higher number, as in 35 and **37**. In the end, both upper and lower lists get an equal share but in an absolutely irregular fashion.

There are also occasions when both numbers of the pair are prime, as in **5** and **7, 11** and **13,** and **101** and **103.** Such pairs are called 'prime twins' and they can be found as far as the list of numbers has been investigated for primes. The density of their occurrence diminishes as the numbers grow larger, just as does the density of the primes themselves. It would seem, however, that the density of prime twins never falls to zero and that the number of prime twins is infinite. That, however, has *never been proved*.

If we consider the numbers of the form $6n+1$ and $6n-1$ only, we find they contain every single prime in existence (except 2 and 3) yet make up only one third of all the numbers in any finite successive list. Is there any way we can translate this into the final-digit business?

The answer is yes!!!! And I use those exclamation points because I come here to something that I am sure has been well known to mathematicians for at least two centuries, but which I have never seen mentioned in any book I have read. I have worked this out independently!

All you have to do is use a six-based system, in which our ordinary numbers look as follows:

10-based: 1, 2, 3, 4, 5, 6, 7, 8, 9, 10, 11, 12, 13, 14, 15, 16, 17 . . .

6-based: 1, 2, 3, 4, 5, 10, 11, 12, 13, 14, 15, 20, 21, 22, 23, 24, 25 . . . (There is no space here to go into details on other-based number systems, but see 'One, Ten, Buckle My Shoe', reprinted in *Adding a Dimension*, Doubleday, 1964.)

In the 6-based system, only numbers ending in the digits one and five could possibly be prime. In the 6-based system we would know at once that 14313234442, 14313234443, 14313234444, and 14313234440 were *not* prime, just by looking at the last digit. On the other hand, 14313234441 and 14313234445 *might* be prime (and, unfortunately, might not).

The point is that in a 6-based system you could instantly eliminate two thirds of the numbers in any finite successive list of numbers just by looking at the final digit, leaving one third to contain all the primes (except 2 and 3). This is better than we can do in the 10-based system, where we eliminate three fifths and leave two fifths.

But what if we use a number as base that does not have two different prime factors as do 6 and 10, but *three* different prime factors? The smallest number which qualifies is $30 = 2 \times 3 \times 5$.

If we use 30 as base, consider that all numbers are of the form $30n$, $30n+1$, $30n+2$, $30n+3$ ... all the way up to $30n+29$. Of these, numbers of the form $30n$, $30n+2$, $30n+4$, and so on, are divisible by 2 and are therefore non-prime; numbers of the form $30n+3$, $30n+9$, $30n+15$, and so on are divisible by 3 and are therefore non-prime; numbers of the form $30n+5$ and $30n+25$ are divisible by 5 and are therefore non-prime. In the end, the only numbers that cannot be divided by 2, 3, or 5 (except for 2, 3, and 5 themselves) and therefore *may* be primes, are numbers of the classes $30n+1$, $30n+7$, $30n+11$, $30n+13$, $30n+17$, $30n+19$, $30n+23$, and $30n+29$.

This sounds like a large number of classes to contain primes, but in the 30-based system there are thirty different digits, one representing every number from 0 to 29 inclusive. And in a 30-based system, numbers ending in twenty-two of these thirty digits are non-prime on the face of it. Only those ending in the eight digits equivalent to our ten-based numbers 1, 7, 11, 13, 17, 19, 23, and 29 *may* be primes.

In the 30-based system, then, we eliminate eleven fifteenths, or $73\frac{1}{3}$ per cent, of any finite successive list of numbers and crowd all the primes (except 2, 3 and 5) into the $26\frac{2}{3}$ per cent remaining.

Of course, you can go still higher. You can use a number system based on 210 (since $210 = 2 \times 3 \times 5 \times 7$) or 2310 (since $2310 = 2 \times 3 \times 5 \times 7 \times 11$) or still higher, going up the scale of multiplied primes as far as you care to go. In each case, you have to leave out of account all the primes that are factors of the number base, but will find all other primes crowded into a smaller and smaller fraction of any finite successive list of numbers.

Here's the way it works as far as I've gone:

| Number Base | % eliminated | % remaining |
|---|---|---|
| 2 | 50 | 50 |
| $2 \times 3 = 6$ | $66\frac{2}{3}$ | $33\frac{1}{3}$ |
| $2 \times 3 \times 5 = 30$ | $73\frac{1}{3}$ | $26\frac{2}{3}$ |
| $2 \times 3 \times 5 \times 7 = 210$ | $77\frac{1}{7}$ | $22\frac{8}{7}$ |

# PRIME QUALITY

I refuse to go higher. You can work it out for 2310 or for any still higher number base yourself.*

Now mind you, the larger you make the number on which you base your number system, the more inconvenient it is to handle that system in practice, regardless of how beautiful it may be in theory. It is perfectly easy to understand the system for writing and handling numbers in a 30-based system, but to try to do so in actual manipulations on paper is a one-way ticket to the booby hatch – at least if your mind is no nimbler than mine.

The gain in prime-concentration in passing to a 30-based system (and I won't even talk about a 210-based system or anything higher) is simply not worth the tremendous loss in manipulability.

Let us therefore stick with the 6-based system, which is not only more efficient as a prime-concentrator than our ordinary 10-based system is, but is actually easier to manipulate once you are used to it.

Or we can put it another way. It is the 6-based system which is, in this respect at least, of prime quality.†

* Since this chapter first appeared, knowledgeable readers have sent me formulas to use in such calculations. If I had known them I would have had a lot less trouble.

† Let there be no groaning in the gallery!

## 10 – EUCLID'S FIFTH

Some of my articles stir up more reader comment than others, and one of the most effective in this respect was one I once wrote in which I listed those who, in my opinion, were scientists of the first magnitude and concluded by working up a personal list of the ten greatest scientists of all time.

Naturally, I received letters arguing for the omission of one or more of my ten best in favor of one or more others, and I still get them, even now, seven and a half years after the article was written.

Usually, I reply by explaining that estimates as to the ten greatest scientists (always excepting the case of Isaac Newton, concerning whom there can be no reasonable disagreement) are largely a subjective matter and cannot really be argued out.

Recently, I received a letter from a reader who argued that Archimedes, one of my ten, ought to be replaced by Euclid, who was not one of my ten. I replied in my usual placating manner, but then went on to say that Euclid was 'merely a systematizer' while Archimedes had made very important advances in physics and mathematics.

But later my conscience grew active. I still adhered to my own opinion of Archimedes taking pride of place over Euclid, but the phase 'merely a systematizer' bothered me. There is nothing necessarily 'mere' about being a systematizer.*

For three centuries before Euclid (who flourished about 300 B.C.) Greek geometers had labored at proving one

---

* Sometimes there is. In all my non-fiction writings I am 'merely' a systematizer. —Just in case you think I'm *never* modest.

geometric theorem or another and a great many had been worked out.

What Euclid did was to make a system out of it all. He began with certain definitions and assumptions and then used them to prove a few theorems. Using those definitions and assumptions plus the few theorems he had already proved, he proved a few additional theorems and so on, and so on.

He was the first, as far as we know, to build an elaborate mathematical system based on the explicit attitude that it was useless to try to prove *everything*; that it was essential to make a beginning with some things that could not be proved but that could be accepted without proof because they satisfied intuition. Such intuitive assumptions, without proof, are called 'axioms'.

This was in itself a great intellectual advance, but Euclid did something more. He picked *good* axioms.

To see what this means, consider that you would want your list of axioms to be complete, that is, they should suffice to prove all the theorems that are useful in the particular field of knowledge being studied. On the other hand they shouldn't be redundant. You don't want to be able to prove all those theorems even after you have omitted one or more of your axioms from the list; or to be able to prove one or more of your axioms by the use of the remaining axioms. Finally, your axioms must be consistent. That is, you do not want to use some axioms to prove that something is so and then use other axioms to prove the same thing to be *not* so.

For two thousand years, Euclid's axiomatic system stood the test. No one ever found it necessary to add another axiom, and no one was ever able to eliminate one or to change it substantially – which is a pretty good testimony to Euclid's judgment.

By the end of the nineteenth century, however, when notions of mathematical rigor had hardened, it was realized that there were many tacit assumptions in the Euclidean system; that is, assumptions that Euclid made without specifically saying that he had made them, and that all his readers also made, apparently without specifically saying so to themselves.

For instance, among his early theorems are several that demonstrate two triangles to the congruent (equal in both shape and size) by a course of proof that asks people to imagine that one triangle is moved in space so that it is superimposed on the other. —That, however, presupposes that a geometrical figure doesn't change in shape and size when it moves. Of course it doesn't, you say. Well, you assume it doesn't and I assume it doesn't and Euclid assumed it doesn't – but Euclid never said he assumed it.

Again, Euclid assumed that a straight line could extend infinitely in both directions – but never said he was making that assumption.

Furthermore, he never considered such important basic properties as the *order* of points in a line, and some of his basic definitions were inadequate—

But never mind. In the last century, Euclidean geometry has been placed on a basis of the utmost rigor and while that meant the system of axioms and definitions was altered, Euclid's geometry remained the same. It just meant that Euclid's axioms and definitions, *plus* his unexpressed assumptions, were adequate to the job.

Let's consider Euclid's axioms now. There were ten of them and he divided them into two groups of five. One group of five was called 'common notions' because they were common to all sciences:

(1) Things which are equal to the same thing are also equal to one another.

(2) If equals are added to equals, the sums are equal.

(3) If equals are subtracted from equals, the remainders are equal.

(4) Things which coincide with one another are equal to one another.

(5) The whole is greater than the part.

These 'common notions' seem so common, indeed so

obvious, so immediately acceptable by intuition, so incapable of contradiction, that they seem to represent absolute truth. They seem something a person could seize upon as soon as he had evolved the light of reason. Without ever sensing the universe in any way, but living only in the luminous darkness of his own mind, he would see that things equal to the same thing are equal to one another and all the rest.

He might then, using Euclid's axioms, work out all the theorems of geometry and, therefore, the basic properties of the universe from first principles, without having observed anything.

The Greeks were so fascinated with this notion that all mathematical knowledge comes from within that they lost one important urge that might have led to the development of experimental science. There were experimenters among the Greeks, notably Ctesibius and Hero, but their work was looked upon by the Greek scholars as a kind of artisanship rather than as science.

In one of Plato's dialogues, Socrates asks a slave certain questions about a geometric diagram and has him answer and prove a theorem in doing so. This was Socrates' method of showing that even an utterly uneducated man could draw truth from out of himself. Nevertheless, it took an extremely sophisticated man, Socrates, to ask the questions, and the slave was by no means uneducated, for merely by having been alive and perceptive for years, he had learned to make many assumptions by observation and example, without either himself or (apparently) Socrates being completely aware of it.

Still as late as 1800, influential philosophers such as Immanuel Kant held that Euclid's axioms represented absolute truth.

But do they? Would anyone question the statement that 'the whole is greater than the part'? Since 10 can be broken up into $6+4$, are we not completely right in assuming that 10 is greater than either 6 or 4? If an astronaut can get into a space capsule, would we not be right in assuming that the volume of the capsule is greater than the volume of the astronaut? How could we doubt the general truth of the axiom?

Well, any list of consecutive numbers can be divided into odd numbers and even numbers, so that we might conclude that in any such list of consecutive numbers, the total of all numbers present must be greater than the total of even numbers. And yet if we consider an *infinite* list of consecutive numbers, it turns out that the total number of all the numbers is equal to the total number of all the even numbers. In what is called 'transfinite mathematics' the particular axiom about the whole being greater than the part simply does not apply.

Again, suppose that two automobiles travel between points A and B by identical routes. The two routes coincide. Are they equal? Not necessarily. The first automobile traveled from A to B, while the second traveled from B to A. In other words, two lines might coincide and yet be unequal since the direction of one might be different from the direction of the other.

Is this just fancy talk? Can a line be said to have direction? Yes, indeed. A line with direction is a 'vector' and in 'vector mathematics' the rules aren't quite the same as in ordinary mathematics and things can coincide without being equal.

In short, then, axioms are *not* examples of absolute truth and it is very likely that there is no such thing as absolute truth at all. The axioms of Euclid are axioms not because they appear as absolute truth out of some inner enlightenment but only because they seem to be true in the context of the real world.

And that is why the geometric theorems derived from Euclid's axioms seem to correspond with what we call reality. They *started* with what we call reality.

It is possible to start with any set of axioms, provided they are not self-contradictory, and work up a system of theorems consistent with those axioms and with each other, even though they are *not* consistent with what we think of as the real world. This does not make the 'arbitrary mathematics' less 'true' than the one starting from Euclid's axioms, only less useful, perhaps. Indeed, an 'arbitrary mathematics' may be *more* useful than ordinary 'common-sense' mathematics in special regions such as those of transfinites or of vectors.

Even so, we must not confuse 'useful' and 'true'. Even if an axiomatic system is so bizarre as to be useful in no conceivable practical sense, we can nevertheless say nothing about its 'truth'. If it is self-consistent that is all we have a right to demand of any system of thought. 'Truth' and 'reality' are theological words, not scientific ones.

But back to Euclid's axioms. So far I have only listed the five 'common notions'. There were also five more axioms on the list that were specifically applicable to geometry and these were later called 'postulates'. The first of these postulates was:

(1) It is possible to draw a straight line from any point to any other point.

This seems eminently acceptable, but are you sure? Can you prove that you can draw a line from the Earth to the Sun? *If* you could somehow stand on the Sun safely and hold the Earth motionless in its orbit, and somehow stretch a string from the Earth to the Sun and pull it absolutely taut, that string would represent a straight line from Earth to Sun. You're sure that this is a reasonable 'thought experiment' and I'm sure it is, too, but we only *assume* that matters can be so. We can't ever demonstrate them, or prove them mathematically.

And, incidentally, what is a straight line? I have just made the assumption that if a string is pulled absolutely taut, it has a shape we would recognize as what we call a straight line. But what is that shape? We simply can't do better than say, 'A straight line is something very, very thin and very, very straight', or, to paraphrase Gertrude Stein, 'A straight line is a straight line is a straight line—'

Euclid defines a straight line as 'a line which lies evenly with the points on itself,' but I would hate to have to try to describe what he means by that statement to a student beginning the study of geometry.

Another definition says that: A straight line is the shortest distance between two points.

But if a string is pulled absolutely taut, it cannot go from the point at one end to the point at the other in any shorter distance, so that to say that a straight line is the shortest distance between two points is the same as saying that it has the shape of an absolutely taut string, and we can still say 'And what shape is that?'

In modern geometry, straight lines are not defined at all. What is said, in essence, is this: Let us call something a line which has the following properties in connection with other undefined terms like 'point', 'plane', 'between', 'continuous', and so on. Then the properties are listed.

Be that as it may, here are the remaining postulates of Euclid:

(2) A finite straight line can be extended continuously in a straight line.

(3) A circle can be described with any point as center and any distance as radius.

(4) All right angles are equal.

(5) If a straight line falling on two straight lines makes the interior angles on the same side less than two right angles, the two straight lines, if produced indefinitely, meet on that side on which are the angles less than the two right angles.

I trust you notice something at once. Of all the ten axioms of Euclid, only one – the fifth postulate – is a long jaw-breaker of a sentence; and only one – the fifth postulate – doesn't make instant sense.

Take any intelligent person who has studied arithmetic and who has heard of straight lines and circles and give him the ten axioms one by one and let him think a moment and he will say, 'Of course!' to each of the first nine. Then recite the fifth postulate and he will surely say, 'What!'

And it will take a long time before he understands what's

going on. In fact, I wouldn't undertake to explain it myself
without a diagram like the one below.

Consider two of the solid lines in the diagram: the one
that runs from point C to point D through point M (call it
line CD after the end points) and the one that runs through
points G, L, and H (line GH). A third line, which runs
through points A, L, M, and B (line AB), crosses both GH
and CD, making angles with both.

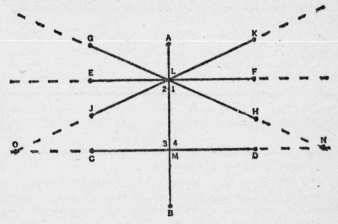

If line CD is supposed to be perfectly horizontal, and line
AB is supposed to be perfectly vertical, then the four angles
made in the crossing of the two lines (angles CMB, BMD,
DML, and LMC) are right angles and are all equal (by
postulate 4). In particular, angles DML and LMC, which
I have numbered in the diagram as 3 and 4, are equal, and
are both right angles.

(I haven't bothered to define 'perfectly horizontal' or 'perfectly vertical' or 'crosses' or to explain why the crossing of a
perfectly horizontal line with a perfectly vertical line produces four right angles, but I am making no pretense of
being completely rigorous. This sort of thing *can* be made
rigorous but only at the expense of a lot more talk than I am
prepared to give.)

Now consider line GH. It is *not* perfectly horizontal. That means the angles it produces at its intersection (I haven't defined 'intersection') with line AB are not right angles and are not all equal. It can be shown that angles ALH and GLB are equal and that angles HLB and GLA are equal but that either of the first pair is not equal to either of the second pair. In particular, angle GLB (labeled 2) is not equal to angle HLB (labeled 1).

Suppose we draw line EF, passing through L, and that line EF is (like line CD) perfectly horizontal. In that case it makes four equal right angles at its intersection with line AB. In particular, angles FLB and ELB are right angles. But angle HLB is contained within angle FLB (what does 'is contained within' mean?) with room to spare. Since angle HLB is only part of FLB and the latter is a right angle then angle HLB (angle 1) is less than a right angle, by the fifth 'common notion'.

In the same way, by comparing angle ELB, known to be a right angle, with angle GLB (angle 2), we can show that angle 2 is greater than a right angle.

The 'interior angles' of the diagram are those on the side of line GH that faces line CD, and those on the side of line CD that faces line GH. In other words, they are angles 1, 2, 3, and 4.

The fifth postulate talks about 'the interior angles on the same side,' that is, 1 and 4 on one side and 2 and 3 on the other. Since we know that 3 and 4 are right angles, that 1 is less than a right angle, and that 2 is more than a right angle, we can say that the interior angles on one side, 1 and 4, have a sum less than two right angles, while the interior angles on the other have a sum greater than two right angles.

The fifth postulate now states that if the lines GH and CD are extended, they will intersect on the side where the interior angles with a sum less than two right angles are located. And, indeed, if you look at the diagram you will see that if lines GH and CD are extended on both sides (dotted lines), they will intersect at point N on the side of interior angles 1 and 4. On the other side, they just move farther and farther apart and clearly will never intersect.

On the other hand, if you draw line JK through L, you would reverse the situation. Angle 2 would be less than a right angle and angle 1 would be greater than a right angle (where angle 2 is now angle JLB and angle 1 is now angle KLB). In that case interior angles 2 and 3 would have a sum less than two right angles and interior angles 1 and 4 would have a sum greater than two right angles. If lines JK and CD were extended (dotted lines), they would intersect at point O on the side of interior angles 2 and 3. On the other side they would merely diverge further and further.

Now that I've explained the fifth postulate at great length (and even then only at the cost of being very un-rigorous) you might be willing to say, 'Oh yes, of course. Certainly! It's obvious!'

Maybe, but if something is obvious, it shouldn't require hundreds of words of explanation. I didn't have to belabor any of the other nine axioms, did I?

Then again, having *explained* the fifth postulate, have I *proved* it? No, I have only interpreted the meaning of the words and then pointed to the diagram and said, 'And indeed, if you look at the diagram, you will see—'

But that's only one diagram. And it deals with a perfectly vertical line crossing two lines of which one is perfectly horizontal. And what if none of the lines are either vertical or horizontal and none of the interior angles are right angles? The fifth postulate applies to *any* line crossing any two lines and I certainly haven't proved that.

I can draw a million diagrams of different types and show that in each specific case the postulate holds, but that is not enough. I must show that it holds in every conceivable case, and this can't be done by diagrams. A diagram can only make the proof clear; the proof itself must be derived by permissible logic from more basic premises already proved, or assumed. This I have not done.

Now let's consider the fifth postulate from the standpoint of moving lines. Suppose line GH is swiveled about L as a pivot in such a way that it comes closer and closer to coinciding with line EF. (Does a straight line remain a straight line while it swivels in this fashion? We can only *assume* it does.) As line GH swivels toward line EF, the point of intersection with line CD (point N) moves farther and farther to the right.

If you started with line JK and swiveled it so that it would eventually coincide with line EF, the intersection point O would move off farther and farther to the left. If you consider the diagram and make a few markings on it (if you have to) you will see this for yourself.

But consider line EF itself. When GH has finally swiveled so as to coincide with line EF, we might say that intersection point N has moved off an infinite distance to the right (whatever we mean by 'infinite distance') and when line JK coincides with line EF, the intersection point O has moved off an infinite distance to the left. Therefore, we can say that line EF and line CD intersect at *two* points, one an infinite distance to the right and one an infinite distance to the left.

Or let us look at it another way. Line EF, being perfectly horizontal, intersects line AB to make four equal right angles. In that case, angles 1, 2, 3, and 4 are *all* right angles and *all* equal. Angles 1 and 4 have a sum equal to two right angles, and so do angles 2 and 3.

But the fifth postulate says the intersection comes on the side where the two interior angles have a sum *less* than two right angles. In the case of lines EF and CD crossed by line AB, neither set of interior angles has a sum less than two right angles and there can be an intersection on neither side.

We have now, by two sets of arguments, demonstrated first that lines EF and CD intersect at two points, each located an infinite distance away, and second that lines EF and CD do not intersect at all. Have we found a contradiction and thus shown that there is something wrong with Euclid's set of axioms?

To avoid a contradiction, we can say that having an intersection at an infinite distance is equivalent to saying there is no intersection. They are different ways of saying the same thing. To agree that 'saying a' is equal to 'saying b' in this case is consistent with all the rest of geometry, so we can get away with it.

Let us now say that two lines, such as EF and CD, which do not intersect with each other when extended any *finite* distance, however great, are 'parallel'.

Clearly, there is only one line passing through L that can be parallel to line CD, and that is line EF. Any line through L that does not coincide with line EF is (however slightly) either of the type of line GH or of line JK, with an interior angle on one side or the other that is less than a right angle. This argument is sleight of hand, and not rigorous, but it allows us to see the point and say: Given a straight line, and a point outside that line, it is possible to draw one and only one straight line through that point parallel to the given line.

This statement is entirely equivalent to Euclid's fifth postulate. If Euclid's fifth postulate is removed and this statement put in its place, the entire structure of Euclidean geometry remains standing without as much as a quiver.

The version of the postulate that refers to parallel lines *sounds* clearer and easier to understand than the way Euclid puts it, because even the beginning student has some notion of what parallel lines look like, whereas he may not have the foggiest idea of what interior angles are. That is why it is in this 'parallel' form that you usually see the postulate in elementary geometry books.

Actually, though, it isn't really simpler and clearer in this form, for as soon as you try to explain what you mean by 'parallel' you're going to run into the matter of interior angles. Or, if you try to avoid that, you'll run into the problem of talking about lines of infinite length, of intersections at an infinite distance being equivalent to no intersection, and that's even worse.

But look, just because I didn't prove the fifth postulate doesn't mean it can't be proven. Perhaps by some line of argument, exceedingly lengthy, subtle and ingenious, it is possible to prove the fifth postulate by use of the other four postulates and the five common notions (or by use of some additional axiom not included in the list which, however, is much simpler and more 'obvious' than the fifth postulate is).

Alas, no. For two thousand years mathematicians have now and then tried to prove the fifth postulate from the other axioms simply because that cursed fifth postulate was so long and so unobvious that it didn't seem possible that it could be an axiom. Well, they always failed and it seems certain they *must* fail. The fifth postulate is just not contained in the other axioms or in any list of axioms useful in geometry and simpler than itself.

It can be argued, in fact, that the fifth postulate is Euclid's greatest achievement. By some remarkable leap of insight, he realized that, given the nine brief and clearly 'obvious' axioms, he could not prove the fifth postulate and that he could not do without it either, and that, therefore, long and complicated though the fifth postulate was, *he had to include it among his assumptions.*

So for two thousand years the fifth postulate stood there: long, ungainly, puzzling. It was like a flaw in perfection, a standing reproach to a line of argument otherwise infinitely stately. It bothered the very devil out of mathematicians.

And then, in 1733, an Italian priest, Girolamo Saccheri, got the most brilliant notion concerning the fifth postulate that anyone had had since the time of Euclid, but wasn't brilliant enough himself to handle it—

Let's go into that in the next chapter.

## 11 – THE PLANE TRUTH

There are occasionally problems in immersing myself in these science essays I write. For instance, I watched a luncheon companion sprinkle salt on his dish after an unsatisfactory forkful, try another bite, and say with satisfaction, 'That's much better.'

I stirred uneasily and said, 'Actually, what you mean is, "I like that much better." In saying merely, "That's much better," you are making the unwarranted assumption that food can be objectively better or worse in taste and the further assumption that your own subjective sensation of taste is a sure guide to the objective situation.'

I think I came within a quarter of an inch of getting that dish, salted to perfection as it was, right in the face; and would have well deserved it, too. The trouble, you see, was that I had just written the previous chapter and was brimful on the subject of assumptions.

So let's get back to that. The subject under consideration is Euclid's 'fifth postulate', which I will repeat here so that you won't have to refer back to it:

If a straight line falling on two straight lines make the interior angles on the same side less than two right angles, the two straight lines, if produced indefinitely, meet on that side on which are the angles less than the two right angles.

All Euclid's other axioms are extremely simple but he apparently realized that this fifth postulate, complicated as it seemed, could not be proved from the other axioms, and must therefore be included as an axiom itself.

For two thousand years after Euclid other geometers kept

trying to prove Euclid too hasty in having given up, and strove to find some ingenious method of proving the fifth postulate from the other axioms, so that it might therefore be removed from the list – if only because it was too long, too complicated, and too not immediately obvious to seem a good axiom.

One system of approaching the problem was to consider the following quadrilateral:

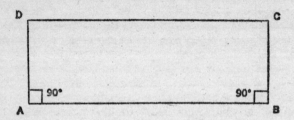

Two of the angles, DAB and ABC are given as right angles in this quadrilateral, and side AD is equal in length to side BC. Given these facts, it is possible to prove that side DC is equal to side AB and that angles ADC and DCB are also right angles (so that the quadrilateral is actually a rectangle) *if* Euclid's fifth postulate is used.

If Euclid's fifth postulate is *not* used, then by using only the other axioms, all one can do is to prove that angles ADC and DCB are equal, but not that they are actually right angles.

The problem then arises whether it is possible to show that from the fact that angles ADC and DCB are equal, it is possible to show that they are also right angles. If one could do that, it would then follow from the fact that quadrilateral ABCD is a rectangle, that the fifth postulate is true. This would have been proven from the other axioms only and it would no longer be necessary to include Euclid's fifth among them.

Such an attempt was first made by the medieval Arabs, who carried on the traditions of Greek geometry while

Western Europe was sunk in darkness. The first to draw this quadrilateral and labor over its right angles was none other than Omar Khayyam (1050–1123).*

Omar pointed out that if angles ADC and DCB were equal, then there were three possibilities: (1) they were each a right angle, (2) they were each less than a right angle, that is 'acute', or (3) they were each more than a right angle, or 'obtuse'.

He then went through a line of argument to show that the acute and obtuse cases were absurd, based on the assumption that two converging lines must intersect.

To be sure, it is perfectly commonsensical to suppose that two converging lines must intersect, but, unfortunately, commonsense or not, that assumption is mathematically equivalent to Euclid's fifth postulate. Omar Khayyam ended, therefore, by 'proving' the fifth postulate by assuming it to be true as one of the conditions of the proof. This is called either 'arguing in a circle' or 'begging the question', but whatever it is called, it is not allowed in mathematics.

Another Arabian mathematician, Nasir Eddin al-Tus (1201–74), made a similar attempt on the quadrilateral, using a different and more complicated assumption to outlaw the acute and obtuse cases. Alas, his assumption was also mathematically equivalent to Euclid's fifth.

Which brings us down to the Italian, Girolamo Saccheri (1667–1733), whom I referred to at the end of the previous chapter and who was both a professor of mathematics at the University of Pisa, and a Jesuit priest.

He knew of Nasir Eddin's work and he, too, tackled the quadrilateral. Saccheri, however, introduced something altogether new, something that in two thousand years no one had thought of doing in connection with Euclid's fifth.

Until then, people had omitted Euclid's fifth to see what would happen, or else had made assumptions that turned

*He wrote clever quatrains which Edward FitzGerald even more cleverly translated into English in 1859, making Omar forever famous as a hedonistic and agnostic poet, but the fact is that he ought to be remembered as a great mathematician and astronomer.

out to be equivalent to Euclid's fifth. What Saccheri did was to begin by assuming Euclid's fifth to be *false*, and to substitute for it some other postulate that was contradictory to it. He planned then to try to build up a geometry based on Euclid's other axioms plus the 'alternative fifth' until he came to a contradiction (proving that a particular theorem was both true *and* false, for instance).

When the contradiction was reached, the 'alternative fifth' would have to be thrown out. If every possible 'alternate fifth' is eliminated in this fashion, then Euclid's fifth must be true. This method of proving a theorem by showing all other possibilities to be absurd is a perfectly mathematical technique* and Saccheri was on the right road.

Working on this system, Saccheri therefore started by assuming that the angles ADC and DCB were both greater than a right angle. With this assumption, plus all the axioms of Euclid *other* than the fifth, he began working his way through what we might call 'obtuse geometry'. Quickly, he came across a contradiction. This meant that obtuse geometry could not be true and that angles ADC and DCB could not each be greater than a right angle.

This accomplishment was so important that the quadrilateral which Omar Khayyam had first used in connection with Euclid's fifth is now called the 'Saccheri quadrilateral'.

Greatly cheered by this, Saccheri then tackled 'acute geometry', beginning with the assumption that angles ADC and DCB were each smaller than a right angle. He must have begun the task lightheartedly, sure that, as in the case of obtuse geometry, he would quickly find a contradiction in acute geometry. If that were so, Euclid's fifth would stand proven and his 'right-angle geometry' would no longer require that uncomfortably long statement as an axiom.

As Saccheri went on from proposition to proposition in his actute geometry, his feeling of pleasure gave way to increasing anxiety, for he did not come across any contra-

* This is equivalent to Sherlock Holmes's famous dictum that when the impossible has been eliminated, whatever remains, however improbable, must be true.

## THE PLANE TRUTH

diction. More and more he found himself faced with the possibility that one could build up a thoroughly self-consistent geometry which was based on at least one axiom that directly contradicted a Euclidean axiom. The result would be a 'non-Euclidean' geometry which might seem against common sense but which would be internally self-consistent and therefore mathematically valid.

For a moment, Saccheri hovered on the very brink of mathematical immortality and – backed away.

He couldn't! To accept the notion of a non-Euclidean geometry took too much courage. So mistakenly had scholars come to confuse Euclidean geometry with absolute truth, that any refutation of Euclid would have roused the deepest stirrings of anxiety in the hearts and minds of Europe's intellectuals. To doubt Euclid was to doubt absolute truth and if there was no absolute truth in Euclid, might it not be quickly deduced that there was no absolute truth anywhere? And since the firmest claim to absolute truth came from religion, might not an attack on Euclid be interpreted as an attack on God?

Saccheri was clearly a mathematician of great potential, but he was also a Jesuit priest and a human being, so his courage failed him and he made the great denial.* When his gradual development of acute geometry went on to the point where he could take it no longer, he argued himself into imagining he had found an inconsistency where, in fact, he hadn't, and with great relief, he concluded that he had proved Euclid's fifth. In 1733, he published a book on his findings entitled (in English) *Euclid Cleared of Every Flaw*, and, in that same year, died.

By his denial Saccheri had lost immortality and chosen oblivion. His book went virtually unnoticed until attention was called to it by a later Italian mathematician, Eugenio Beltrami (1835–1900), *after* Saccheri's failure had been made good by others. Now what we know of Saccheri is just this: that he had his finger on a major mathematical discovery a

* I am not blaming him. Placed in his position, I would undoubtedly have done the same. It's just too bad, that's all.

century before anybody else and had lacked the guts to keep his finger firmly on it.

Let us move forward nearly a century to the German mathematician Karl F. Gauss (1777–1855). It can easily be argued that Gauss was the greatest mathematician who ever lived. Even as a young man he astonished Europe and the scientific world with his brilliance.

He considered Euclid's fifth about 1815 and came to the same conclusion to which Euclid had come – that the fifth *had* to be made an axiom because it *couldn't* be proved from the other axioms. Gauss further came to the conclusion from which Saccheri had shrunk away – that there were other self-consistent geometries which were non-Euclidean, in that an alternate axiom replaced the fifth.

And then *he* lacked the guts to publish, too, and here I disclaim sympathy. The situation was different. Gauss had infinitely more reputation than Saccheri; Gauss was not a priest; Gauss lived in a land where, and at a time when, the hold of the Church was less to be feared. Gauss, genius or not, was just a coward.

Which brings us to the Russian mathematician Nikolai Ivanovich Lobachevski (1793–1856).* In 1826, Lobachevski also began to wonder if a geometry might not be non-Euclidean and yet consistent. With that in mind, he worked out the theorems of 'acute geometry' as Saccheri had done a century earlier, but in 1829, Lobachevski did what neither Saccheri nor Gauss had done. He did *not* back away and he *did* publish. Unfortunately, what he published was an article in Russian called 'On the Principles of Geometry' in a local periodical (he worked at the University of Kazan, deep in provincial Russia).

Who reads Russian? Lobachevski remained largely unknown. It wasn't until 1840 that he published his work in

* Nikolai Ivanovich Lobachevski is mentioned in one of Tom Lehrer's satiric songs and to any Tom Lehrer fan (like myself) it seems strange to see the name mentioned in a serious connection, but Lehrer is a mathematician by trade and he made use of a real name.

German and brought himself to the attention of the world of mathematics generally.

Meanwhile, though, a Hungarian mathematician, János Bolyai (1802–60), was doing much the same thing. Bolyai is one of the most romantic figures in the history of mathematics since he also specialized in such things as the violin and the dueling sword – in the true tradition of the Hungarian aristocrat. There is a story that he once fenced with thirteen swordsmen one after the other, vanquishing them all – and playing the violin between bouts.

In 1831, Bolyai's father published a book on mathematics. Young Bolyai had been pondering over Euclid's fifth for a number of years and now he persuaded his father to include a twenty-six-page appendix in which the principles of acute geometry were described. It was two years after Lobachevski had published but as yet no one had heard of the Russian and nowadays, Lobachevski and Bolyai generally share the credit for having discovered non-Euclidean geometry.

Since the Bolyais published in German, Gauss was at once aware of the material. His commendation would have meant a great deal to the young Bolyai. Gauss still lacked the courage to put his approval into print, but he did praise Bolyai's work verbally. And then, he couldn't resist – He told Bolyai he had had the same ideas years before but hadn't published, and showed him the work.

Gauss didn't have to do that. His reputation was unshakable; even without non-Euclidean geometry, he had done enough for a dozen mathematicians. Since he had lacked the courage to publish, he might have had the decency to let Bolyai take the credit. But he didn't. Genius or not, Gauss was a mean man in some ways.

Poor Bolyai was so embarrassed and humiliated by Gauss's disclosure, that he never did any further work in mathematics.

And what about obtuse geometry? Saccheri had investigated that and found himself enmeshed in contradiction, so that had been thrown out. Still, once the validity of non-Euclidean geometry had been established, was there no way of rehabilitating obtuse geometry, too?

Yes, there was – but only at the cost of making a still more radical break with Euclid. Saccheri, in investigating obtuse geometry had made use of an unspoken assumption that Euclid himself had also used – that a line could be infinite in length. This assumption introduced no contradiction in acute geometry or in right-angle geometry (Euclid's), but it did create trouble in obtuse geometry.

But then, drop that too. Suppose that, regardless of 'common sense' you were to make the assumption that any line had to have some maximum finite length. In that case all the contradiction in obtuse geometry disappeared and there was a second valid variety of non-Euclidean geometry. This was first shown in 1854 by the German mathematician Georg F. Riemann (1826–66).

So now we have three kinds of geometry, which we can distinguish by using statements that are equivalent to the variety of fifth postulate used in each case:

(A) Acute geometry (non-Euclidean): Through a point not on a given line, an infinite number of lines parallel to the given line may be drawn.

(B) Right-angle geometry (Euclidean): Through a point not on a given line, one and only one line parallel to the given line may be drawn.

(C) Obtuse geometry (non-Euclidean): Through a point not on a given line, no lines parallel to the given line may be drawn.

You can make the distinction in another and equivalent way:

(A) Acute geometry (non-Euclidean): The angles of a triangle have a sum less than 180°.

(B) Right-angle geometry (Euclidean): The angles of a triangle have a sum exactly equal to 180°.

(C) Obtuse geometry (non-Euclidean): The angles of a triangle have a sum greater than 180°.

You may now ask: But which geometry is *true*?

If we define 'true' as internally self-consistent, then all three geometries are equally true.

Of course, they are inconsistent with each other and perhaps only one corresponds with reality. We might therefore ask: Which geometry corresponds to the properties of the real universe?

The answer is, again, that all do.

Let us, for instance, consider the problem of traveling from point A on Earth's surface to point B on Earth's surface, and suppose we want to go from A to B in such a way as to traverse the least distance.

In order to simplify the results, let us make two assumptions. First, let us assume that the Earth is a perfectly smooth sphere. This is almost true, as a matter of fact, and we can eliminate mountains and valleys and even the equatorial bulge without too much distortion.

Second, let us assume that we are confined in our travels to the surface of the sphere and cannot, for instance, burrow into its depth.

In order to determine the shortest distance from A to B on the surface of the Earth, we might stretch a thread from one point to the other and pull it taut. If we were to do this between two points on a plane, that is, on a surface like that of a flat blackboard extended infinitely in all directions, the result would be what we ordinarily call a 'straight line'.

On the surface of the sphere, the result, however, is a curve, and yet that curve is the analogue of a straight line, since that curve is the shortest distance between two points on the surface of a sphere. There is difficulty in forcing ourselves to accept a curve as analogous to a straight line because we've been thinking 'straight' all our lives. Let us use a different word, then. Let us call the shortest distance between two points on any given surface a 'geodesic'.*

On a plane, a geodesic is a straight line; on a sphere, a

* 'Geodesic' is from Greek words meaning 'to divide the Earth' because any geodesic on the face of the Earth, if extended as far as possible, divides the surface of the Earth into two equal parts.

geodesic is a curve, and, specifically, the arc of a 'great circle'. Such a great circle has a length equal to the circumference of the sphere and lies in a plane that passes through the center of the sphere. On the Earth, the equator is an example of a great circle and so are all the meridians. There are an infinite number of great circles that can be drawn on the surface of any sphere. If you choose any pair of points on a sphere and connect each pair by a thread which is pulled taut, you have in each case the arc of a different great circle.

You can see that on the surface of a sphere, there is no such thing as a geodesic of infinite length. If it is extended, it simply meets itself as it goes around the sphere and becomes a closed curve. On the surface of the Earth, a geodesic can can be no longer than 25,000 miles.

Furthermore, any two geodesics drawn on a sphere intersect if produced indefinitely, and do so at two points. On the surface of the Earth, for instance, any two meridians meet at the north pole and the south pole. This means that, on the surface of a sphere, through any point not on a given geodesic, no geodesic can be drawn parallel to the given geodesic. No geodesic can be drawn through the point that won't sooner or later intersect the given geodesic.

Then, too, if you draw a triangle on the surface of a sphere, with each side the arc of a great circle, then the angles will have a sum greater than 180°. If you own a globe, imagine a triangle with one of its vertices at the north pole, with a second at the equator and 10° west longitude, and the third at the equator and 100° west longitude. You will find that you will have an equilateral triangle with each one of its angles equal to 90°. The sum of the three angles is 270°.

This is precisely the geometry that Riemann worked out, if the geodesics are considered, the analogues of straight lines. It is a geometry of finite lines, no parallels, and triangular angle-sums greater than 180°. What we have been calling 'obtuse geometry' then might also be called 'sphere geometry'. And what we have been calling 'right-angle geometry' or 'Euclidean geometry' might also be called 'plane geometry'.

In 1865, Eugenio Beltrami drew attention to a shape called a 'pseudosphere', which looks like two cornets joined wide mouth to wide mouth, and with each cornet extending infinitely out in either direction, ever narrowing but never quite closing. The geodesics drawn on the surface of a pseudosphere fulfill the requirements of acute geometry.

Geodesics on a pseudosphere are infinitely long and it is possible for two particular geodesics to be extended indefinitely without intersecting and therefore to be parallel. In fact, it is possible to draw two geodesics on the surface of a pseudosphere that *do* intersect and yet have neither one intersecting a third geodesic lying outside the two.* In fact, since an infinite number of geodesics can be drawn in between the two intersecting geodesics, all intersecting in the same point, there are an infinite number of possible geodesics through a point, all of which are parallel to another geodesic not passing through the point.

In other words 'acute geometry' can be looked at as 'pseudosphere geometry'.

But now – granted that all three geometries are equally valid under circumstances suiting each – which is the best description of the universe as a whole?

This is not always easy to tell. If you draw a triangle with geodesics of a given length on a small sphere and then again on a large sphere, the sum of the angles of the triangle will be greater than 180° in either case, but the amount by which it is greater will be greater in the case of the small sphere.

If you imagine a sphere growing larger and larger, a triangle of a given size on its surface will have an angle-sum

---

* This sounds nonsensical because we are used to thinking in terms of planes where the geodesics are straight lines and where two intersecting lines cannot possibly be both parallel to a third line. On a pseudosphere, the geodesics curve, and curve in such a way as to make the two parallels possible.

closer and closer to 180° and eventually even the most refined possible measurement won't detect the difference. In short, a small section of a very large sphere is almost as flat as a plane and it becomes impossible to tell the difference.

This is true of the Earth, for instance. It is because the Earth is so large a sphere that small parts of it look flat and that it took so long for mankind to satisfy himself that it was spherical despite the fact that it looked flat.

Well, there is a similar problem in connection with the universe generally.

Light travels from point to point in space; from the Sun to the Earth, or from one distant Galaxy to another, over distances many times those possible on Earth's surface.

We assume that light in traveling across the parsecs travels in a straight line but, of course, it really travels in a geodesic, which may or may not be a straight line. If the universe obeys Euclidean geometry, the geodesic is a straight line. If the universe obeys some non-Euclidean geometry, then the geodesics are curves of one sort or another.

It occurred to Gauss to form triangles with beams of light traveling through space from one mountaintop to another, and measure the sum of the angles so obtained. To be sure, the sums turned out to be just about 180°, but were they *exactly* 180°? That was impossible to tell. If the universe were a sphere millions of light-years in diameter and if the light beams followed the curvings of such a sphere, no conceivable direct measurement possible today could detect the tiny amount by which the angle sum exceeded 180°.

In 1916, however, Einstein worked out the General Theory of Relativity, and found that in order to explain the workings of gravitation, he had to assume a universe in which light (and everything else) traveled in non-Euclidean geodesics.

By Einstein's theory, the universe is non-Euclidean and is, in fact, an example of 'obtuse geometry'.

To put it briefly, then, Euclidean geometry, far from being the absolute and eternal verity it was assumed to be for two thousand years, is only the highly restricted and abstract geometry of the plane, and one that is merely an approximation of the geometry of such important things as the universe and the Earth's surface.

It is not the plain truth so many have taken for granted it was – but only the plane truth.*

* Well, *I* think it's clever.

*D – The Problem of the Platypus*

# 12 – HOLES IN THE HEAD

A friend said to me once that he would love to see my filing system. So I took him to my office and said, 'This file is for correspondence. Here I keep old manuscripts. Here I have manuscripts in preparation. This is the card file of my books – of my shorter fiction – of my shorter non-fiction—'

'No, no, no,' he said. 'That's all trivial. Where do you keep your reference files?'

'What reference files?' said I, blankly. (I very often say things blankly. I think it's part of my charm – or maybe naïveté.)

'The cards on which you list items you may need for future articles or books and then file them according to various subjects.'

'I don't do that,' I said, growing anxious. 'Am I supposed to?'

'But how do you keep things straight in your head, then?'

I was glad to be able to answer that one definitely. 'I don't know,' I said, and he seemed pretty annoyed with me.

Well, I *don't*. All I know is that I've been a classifier ever since I can remember. Everything falls into categories with me. Everything is divided and counted up and put into neat stacks in my mind. I don't worry about it; it happens by itself.

Of course, I sometimes worry about the details. For instance, what with one thing or another, the actual number of books I have published has become an issue. I am forever being asked, 'How many books have you published?'*

But what's a book?

* The answer is 117 at the moment of writing, if you are dying of curiosity.

Recently, the second edition of my book *The Universe* was published. Do I count that as a new book? Of course not. It's updated but the updating doesn't represent enough in the way of change to make me consider the book 'new'. On the other hand, later, the third edition of my book *The Intelligent Man's Guide To Science* is being published. I counted the second edition as a new book and I intend to count the third edition as such, because in each case the changes introduced were substantial and as time- and energy-consuming as a new book would have been.

You might think all this is something I can chop and change to please myself, but not exactly. In my book *Opus 100*, I listed my first hundred books in chronological order and that list became 'official.' But is it correct? Was I right in omitting this or that item from the list or, for that matter, including this item or that?

Unimportant? Sure, but it does help me sympathize with those classifiers who involve themselves with more intricate matters than a listing of books. For instance—

How do you tell a mammal from a reptile?

The easiest and quickest way is to decide that a mammal is covered with hair and a reptile is covered with scales. Of course, you have to be liberal in making this distinction. Some organisms we consider mammals don't have very much hair. Human beings don't – but we have *some* hair. Elephants have even less, but they have some. Whales have still less, but even they have some. Dolphins usually have anywhere from two to eight hairs near the tip of the snout. Even in those whales where hair is altogether absent, it is present at some time in the fetal development.

And one hair is, in this respect, as good as a million, for any hair at all is the hallmark of the mammal. No creature that we consider to be definitely a non-mammal has even one true hair. They may have structures that look like hair, but the resemblance disappears if we consider its microscopic structure, its chemical makeup, its anatomical origin, or all three.

A somewhat less useful distinction is that mammals (well, most) bring forth their young alive, while reptiles (well, most) don't. Some reptiles, such as sea snakes, bring forth living young, but in doing so they merely retain the eggs within their body till they hatch. The developing embryos find their food within the egg and the fact that the egg is within the body is a point for security, but not for nourishment.

Mammals, on the other hand, or most of them, feed the developing young out of the maternal bloodstream by means of an organ called the 'placenta', in which the mother's blood vessels and the blood vessels of the embryo come close enough to allow molecules to seep across: food from mother to embryo, wastes from embryo to mother. (There is no actual joining of bloodstreams, however.)

A minority of mammals bring forth living, but very poorly developed, young, and these must then continue their development in a special maternal pouch outside the body. A still smaller minority of mammals lay eggs. But even the egg-layers have hair.

Another point is that mammals feed their newborn young on milk secreted by special maternal glands. This is true even of the non-placental mammals; even of the egg-layers. And this is *not* true for any animal without hair (not one!). Milk seems to be a purely mammalian product and it is this, more than anything else, which seems to have impressed the classifiers. The very word 'mammal' is from the Latin *mamma*, meaning 'breast'.

Then, too, mammals maintain a constant internal temperature even though the environmental temperature may vary widely. Reptiles, on the other hand, have an internal temperature that tends, more or less, to match that of the environment. Since the internal temperature of mammals is close to 100°F. and is therefore generally higher than the environmental temperature, mammals feel warm to the touch while reptiles feel, by comparison, cold. That is why we speak of mammals as 'warm-blooded' and reptiles as 'cold-blooded', missing the essential point that the internal temperature is constant in the former case and inconstant in the latter.

(To be sure, birds are warm-blooded, too, but there is no danger in confusing a bird and a mammal. All birds, without exception, have feathers and all non-birds, without exception, do not have feathers. —And except for birds and mammals, all organisms are cold-blooded.)

I have by no means listed all the differences between mammals and reptiles, only those that the non-biologist can tell by looking at the creatures from a distance. If we want to indulge in dissection, we can discover others. For instance, mammals have a flat muscle called the 'diaphragm' which divides the chest from the abdomen. The diaphragm, as it contracts, increases the volume of the chest cavity (at the expense of the abdominal cavity, which doesn't care) and helps draw air into the lungs. Reptiles do not have a diaphragm. In fact, no non-hairy organism does.

So far, so good. But now we pass on to extinct creatures which biologists can study only in fossil form. Paleontologists (those biologists specializing in extinct species) have no hesitation in looking at a fossil and saying that it is reptilian or that it is mammalian. The question at once arises: How?

All the really obvious distinctions can't be used since, in general, all that the fossils offer us are the remains of what used to be bones and teeth. You can't look at a handful of bones and teeth and find traces of hair or breasts or milk or placentae or diaphragms.

All you can do is compare the bones and teeth with those of modern reptiles and mammals and see if there are strictly hard-tissue distinctions. Then, you might assume that if an extinct creature had bones characteristic of mammals, it must also have had hair, breasts, a diaphragm, and the rest.

Consider the skull. In the most primitive and earliest reptiles, the skull behind the eyes was solid bone and on the other side of the bone were the jaw muscles. There was a tendency, however, to expose the jaw muscles and give them freer play, so that many reptiles had openings in the skull bounded by bony arches. The loss in sheer defensive strength was more than made up for by the improvement in the offense represented by large, stronger jaws that could go snap! more firmly. On the balance, then, the reptiles which happened to develop those openings passed on to greater things.

(Yet evolutionary 'advances' are never universal, and never the only answer. One group of reptiles that had no use for a hole in the head, managed to survive for hundreds of millions of years and flourish, after a fashion, even today, though many, many hole-in-the-head groups have vanished. I'm talking about the turtles, whose jaw muscles are hidden under a solid wall of bone.)

The reptiles developed openings in their skulls in a variety of patterns and, indeed, are classified into groups according to those patterns. This is not because this pattern is of overwhelming physiological importance in itself, but only because it is convenient, since if you have any part at all of a reptile, however long dead, you are likely to have its skull.

But what about mammals, which are descended from reptiles? They have a single opening on either side of the skull just behind the eye bounded on the bottom by a narrow bony arch called the 'zygomatic arch'.

So the paleontologist can look at a skull and from the nature of the openings tell at once whether it is reptilian or mammalian.

Then, again, the lower jaw of a reptile is made up of seven different bones, fused tightly into a strong structure. The lower jaw of a mammal is a single bone. (Some of the missing bones developed into the tiny bones of the middle ear. This is not as strange as it sounds. If you put your finger at the point where lower jaw meets upper jaw and where the old reptilian bones existed, you will find you are not very far from your ear.)

As for the teeth, those of reptiles tended to be undifferentiated and all alike, of conelike structure. In mammals, the teeth are highly differentiated, cutting incisors in front, grinding molars in back, with tearing canines and premolars between.

Since mammals evolved from reptilian forebears is there any way of recognizing which group of reptiles possessed the distinction of being our ancestors? Certainly no living group of reptiles seems to be descended from anything mammalian or even approaching the mammalian. We must look for some group that left no reptilian descendants at all.

One such group, now entirely extinct (as reptiles), is called the 'Synapsida'. These had a single skull opening on either side of the head and included members who showed the clear beginnings of mammalism.

There were two important groups of synapsids. The earlier, dating back some three hundred million years, were members of the order 'Pelycosauria'. The pelycosaurs are interesting chiefly because their skulls seem to show the beginning of a zygomatic arch. Furthermore, their teeth show some differentiation. The front teeth are incisorlike and behind them are teeth that are rather like canines. There are no molars however. The rear teeth are reptilian cones.

After flourishing for fifty million years or so, the pelycosaurs gave way to a group of synapsids of the order 'Therapsida'. Undoubtedly, the therapsids were descended from a particular species of pelycosaurs.

The therapsids are clearly further on the road to mammalism than any of the pelycosaurs were. The zygomatic arch is much more mammal-like among them than among the pelycosaurs; so much so, in fact, that the feature gives them their name. 'Therapsida' is from Greek words meaning 'beast opening'. In other words, the opening in the skull is beastlike where 'beast' is the common term for what zoologists would call mammals.

Further, the teeth are much more differentiated among the therapsids than among the pelycosaurs. A well-known therapsid which lived about 220 million years ago in South Africa had a skull and teeth that were so doglike that it is called 'Cynognathus' (dog jaw'). The back teeth of Cynognathus are even beginning to look like molars.

What's more, while the chin of the therapsids was made up of seven bones, in typical reptilian fashion, the center bone or 'dentary', was by far the largest. The other six bones, three on each side, were crowded toward the joint of the lower jaw with the upper – on their way to the ear, so to speak.

In another respect, too, the therapsids showed a 'progressive' feature. (We tend to call 'progressive' anything that seems to move in the direction of ourselves.) Early reptiles, including the pelycosaurs, tended to have their legs splayed out so that the upper part, above the knee, was horizontal. This is a rather inefficient way of suspending the weight of the body.

Not so the therapsids. In their case, the legs were drawn beneath the body, with the upper parts, as well as the lower, tending to be vertical. This makes for better support, allows faster movement with less energy expenditure, and is a typically mammalian characteristic. Apparently, the superior efficiency of the vertical leg meant there was no virtue in particularly long toes. Primitive reptiles tended to have four or even five joints in their middle toes. The therapsids, however, had two joints in the first toes, and three joints in the others. Again, this is the way it is in mammals.

The therapsids, however, did not endure. While we may root for them as our great-ever-so-great-grandfathers, the fact is that about two hundred million years ago, the 'archosaurs', the creatures representing what we loosely call dinosaurs were coming into their own. As they (no ancestors of ours) rapidly grew in size and specialization, they crowded out the therapsids.

By 150 million years ago, the therapsids were clean gone forever, every single one of them extinct.

Well, not really! Some small therapsids remained, but they had grown so mammal-like, as nearly as we can tell from the very few fossil remnants left behind, that we don't call them therapsids any more. We call them mammals.

After the mammals came on the scene, they managed to survive through a hundred million years or so of archosaurian dominance. Then, after the archosaurs vanished, about seventy million years ago, the mammals continued to survive and burst into a flood of differentiation and specialization that made this latest period of Earth's existence the 'age of the mammals'.

The question now is: Why did the mammals survive when the therapsids generally did not? The archosaurs proved utterly superior to the therapsids; why not to those therapsidian offshoots, the mammals, as well? It couldn't have been that the mammals were particularly brainy, because primitive mammals aren't. They are not very brainy even today, much less a hundred million years ago.

Nor could it be because of their advanced reproductive system, the bearing of live young, for instance. The development of a placenta, or even of a pouch, did not take place till near the end of the archosaurian dominance. For nearly a hundred million years the mammals survived as egg-layers.

It couldn't have been their advanced teeth or legs or anything skeletal that the therapsids had, generally, for none of that helped the therapsids, generally.

Actually, the best guess is that the trick of survival was warm-bloodedness, the development of a constant internal temperature. The control of the internal temperature meant that a mammal could withstand the direct rays of a hot sun much more easily than a reptile could. It meant that a mammal was warm and agile on a cold morning when reptiles were cold, stiff, and sluggish.

If a mammal carefully confined his activity to the chilly hours or if he were trapped by a reptile in the heat and could escape by darting into the hot sun – it would tend to survive. But for mammals to have survived in this fashion, their warm-bloodedness must have been well developed from the start and that couldn't happen overnight.

We might conclude, then, that in addition to those changes in the therapsids that we can see in the skeleton, there must have been additional changes that made warm-bloodedness possible. The mammals survived because of all the therapsids, warm-bloodedness had developed most efficiently among them.

Are there any signs of the beginnings of such changes among the reptilian precursors of the mammals? Well, a number of species of pelycosaurs had long bony processes to their vertebrae that thrust high into the air. Apparently, skin stretched across these processes so that pelycosaurs possessed a high, ribbed 'sail'.

Why? The American zoologist Alfred Sherwood Romer has suggested it was an air-conditioning device (like the huge fanlike ears of the African elephant). Heat is gained or lost through the surface of the body and the pelycosaurian sail can easily double the surface area available. On a cool morning, the sail will pick up the Sun's heat and warm the creature much more quickly than would be the case for a similar organism without a sail. Again, on a hot day, a pelycosaur could stay in the shade and lose heat rapidly through the blood vessels engorging the sail.

The sail, in short, served to make the pelycosaur's internal temperature more nearly constant than was the case in other similar reptiles. Their therapsid descendants had no sails, however, and it couldn't be that they had abandoned temperature control, since their descendants, the mammals, had it in such superlative degree.

It must be that the therapsids had developed something better than the sail. A high metabolic rate to produce heat in greater quantity might be developed and then hair (which is only modified scales) to serve as an insulating device that would cut down heat loss on cold days. They might also develop sweat glands to get rid of heat on hot days in more efficient manner than by means of a sail.

In short, could the therapsids have been hairy and sweaty, as mammals are? We can never tell from the fossils.

And did those species which best developed hairiness and sweatiness become what we call mammals and did they survive where the less advanced other therapsids did not?

Let's look in another direction. In reptiles, the nostrils open into the mouth just behind the teeth. This means that reptiles can breathe with their mouths closed – and empty. When the mouth is full, breathing stops. In the case of the cold-blooded reptiles, not much harm is done. The reptilian need for oxygen is relatively low and if the supply is cut off temporarily during eating, so what?

Mammals, however, have to maintain a high metabolic rate at all times if they are to be warm-blooded, and that means that the oxidation of foodstuffs (from which heat is obtained) must continue steadily. The oxygen supply must not be cut off for more than a couple of minutes at any time. This is made possible by the fact that mammals have a palate, a roof to the mouth. When they breathe, air is led above the mouth to the throat. It is only when they are actually in the act of swallowing that the breath is cut off and this is a matter of a couple of seconds only.

It is interesting, then, that a number of late therapsid species had developed a palate. This might be taken as a pretty good indication that they were warm-blooded.

It would seem then that if we could see therapsids in their living state and not as a handful of stony bones, we would see hairy, sweaty creatures that we might easily mistake for mammals. We might then wonder which hairy, sweaty creatures were reptiles and which were mammals. How would we draw the line?

Nowadays, it might seem, the problem is not a crucial one. All the hairy warm-blooded creatures in existence are called mammals. —And yet, are we justified in doing so?

In the case of the placentals and the marsupials, we are surely justified. They developed their placentas and their pouches about eighty million years ago, after the mammals had already existed for some hundred million years. The early mammals must have been egg-layers and so, therefore, must have been their therapsid forebears. If we want to look for the boundary line between therapsids and mammals, we must therefore look among the hairy egg-layers.

As it happens, there are still six species of such hairy egg-layers alive today, existing only in Australia, Tasmania, and New Guinea, islands that split off from Asia before the more efficient placental mammals developed, so that the egg-layers were spared what would otherwise have been a fatal competition. The egg-layers were first discovered in 1792 and for a while biologists found it hard to believe they could really exist. It took a long time before they got over suspecting a hoax – hairy creatures that laid eggs seemed a contradiction in terms.

The best-known of the egg-layers is the 'duckbill platypus' (the last part of the name means 'flat-foot' and the first part refers to the horny sheath on its nose that looks like a duck's bill). It is also called 'Ornithorhynchus' from Greek words meaning 'bird beak'.

These egg-layers have hair, of course, perfectly good hair, but so (very likely) had at least some therapsids. The egg-layers also produce milk, although their mammary glands have no nipples and the young must lick the hair where the milk oozes out. However, some therapsid species might also have produced milk in that fashion. We can't tell from the bones.

In some respects, the egg-layers lean strongly toward the side of the reptiles. Their body temperature is much less perfectly controlled than that of other mammals and some of them possess venom. The platypus, for instance, has a horny spur at each ankle which secretes venom; and though a number of reptiles are venomous, no mammals (other than the egg-layers) are.

Then, too, *because* they are egg-layers, they have a single abdominal opening, a 'cloaca', which serves as a common passageway for urine, feces, eggs, and sperm. All living birds and reptiles (also egg-layers) possess cloacae, but no mammals, other than those few egg-layers, do. For this reason, the egg-layers are called 'monotremes' ('one-hole').

To most zoologists, the hair and the milk spell mammal unmistakably, but the eggs, the cloaca, and the venom are sufficiently reptilian so that the egg-layers are placed in a subclass 'Prototheria' ('first beasts') while all other mammals, marsupials and placentals alike, are in the subclass 'Theria' ('beasts').

The question arises, though: Are the monotremes really the first of the mammals, or are they rather the last of the therapsids? Are they really reptiles that have the outer appearance of mammals, as did, perhaps, a number of late therapsid species; or are they mammals that have retained some reptilian characteristics?

This may sound like a purely semantic matter, but zoologists must make decisions in such matters and, if possible, come to agreement over it.

An American zoologist, Giles T. MacIntyre, has recently entered the fray, using skeletal characteristics as the criterion. (We have only the skeleton as direct evidence in the therapsid case.) He has concentrated on the region near the ear, where some of the reptilian jawbones became mammalian ear bones and where you might expect some useful distinction between the two classes.

There is a 'trigeminal nerve' which leads from the jaw muscles to the brain. In all reptiles, without exception, it passes through a little hole in the skull that lies between two particular bones that make up the skull. In all marsupial and placental mammals, without exception, it passes through a little hole that pierces *through* one of the skull bones.

Then let us forget about hair and milk and eggs and warm-bloodedness, and reduce it to a matter of holes in the head. Does the trigeminal nerve of the monotremes pass through a skull bone or between two skull bones? The answer has been: *Through* a skull bone.

That would mean the monotremes are mammals.

Not so, says MacIntyre. The study of the trigeminal nerve was made in adult monotremes, where the skull bones are fused and the boundaries hard to make out. In young monotremes, the skull bones are not as well developed and are more clearly separated (as is true of young mammals generally). In young monotremes, MacIntyre says, it is clear that the trigeminal nerve goes between two bones and it is only in the adult skull that bone fusions obscure the fact.

If MacIntyre is correct, we may therefore say that the therapsids never became entirely extinct and that the monotremes represent living therapsids, living reptiles so similar to mammals in some ways as to have been considered mammals for nearly two centuries.

Does this matter to anyone but a few zoologists?

Well, it matters to me. Emotionally, I'm all the way on MacIntyre's side. I want the therapsids to have survived!

*E – The Problem of History*

# 13 – THE EUREKA PHENOMENON

In the old days, when I was writing a great deal of fiction, there would come, once in a while, moments when I was stymied. Suddenly, I would find I had written myself into a hole and could see no way out. To take care of that, I developed a technique which invariably worked.

It was simply this – I went to the movies. Not just any movie. I had to pick a movie which was loaded with action but which made no demands on the intellect. As I watched, I did my best to avoid any conscious thinking concerning my problem, and when I came out of the movie I knew exactly what I would have to do to put the story back on the track.

It never failed.

In fact, when I was working on my doctoral dissertation, too many years ago, I suddenly came across a flaw in my logic that I had not noticed before and that knocked out everything I had done. In utter panic, I made my way to a Bob Hope movie – and came out with the necessary change in point of view.

It is my belief, you see, that thinking is a double phenomenon, like breathing.

You can control breathing by deliberate voluntary action: you can breathe deeply and quickly, or you can hold your breath altogether, regardless of the body's needs at the time. This, however, doesn't work well for very long. Your chest muscles grow tired, your body clamors for more oxygen, or less, and you relax. The automatic involuntary control of breathing takes over, adjusts it to the body's needs, and unless you have some respiratory disorder, you can forget about the whole thing.

Well, you can think by deliberate voluntary action, too,

and I don't think it is much more efficient on the whole than voluntary breath control is. You can deliberately force your mind through channels of deductions and associations in search of a solution to some problem and before long you have dug mental furrows for yourself and find yourself circling round and round the same limited pathways. If those pathways yield no solution, no amount of further conscious thought will help.

On the other hand, if you let go, then the thinking process comes under automatic involuntary control and is more apt to take new pathways and make erratic associations you would not think of consciously. The solution will then come while you *think* you are *not* thinking.

The trouble is, though, that conscious thought involves no muscular action and so there is no sensation of physical weariness that would force you to quit. What's more, the panic of necessity tends to force you to go on uselessly, with each added bit of useless effort adding to the panic in a vicious cycle.

It is my feeling that it helps to relax, deliberately, by subjecting your mind to material complicated enough to occupy the voluntary faculty of thought, but superficial enough not to engage the deeper involuntary one. In my case, it is an action movie; in your case, it might be something else.

I suspect it is the involuntary faculty of thought that gives rise to what we call 'a flash of intuition', something that I imagine must be merely the result of unnoticed thinking.

Perhaps the most famous flash of intuition in the history of science took place in the city of Syracuse in third-century B.C. Sicily. Bear with me and I will tell you the story—

About 250 B.C., the city of Syracuse was experiencing a kind of Golden Age. It was under the protection of the rising power of Rome, but it retained a king of its own and considerable self-government; it was prosperous; and it had a flourishing intellectual life.

The king was Hieron II, and he had commissioned a new golden crown from a goldsmith, to whom he had given an ingot of gold as raw material. Hieron, being a practical man, had carefully weighed the ingot and then weighed the crown he received back. The two weights were precisely equal. Good deal!

But then he sat and thought for a while. Suppose the goldsmith had subtracted a little bit of the gold, not too much, and had substituted an equal weight of the considerably less valuable copper. The resulting alloy would still have the appearance of pure gold, but the goldsmith would be plus a quantity of gold over and above his fee. He would be buying gold with copper, so to speak, and Hieron would be neatly cheated.

Hieron didn't like the thought of being cheated any more than you or I would, but he didn't know how to find out for sure if he had been. He could scarcely punish the goldsmith on mere suspicion. What to do?

Fortunately, Hieron had an advantage few rulers in the history of the world could boast. He had a relative of considerable talent. The relative was named Archimedes and he probably had the greatest intellect the world was to see prior to the birth of Newton.

Archimedes was called in and was posed the problem. He had to determine whether the crown Hieron showed him was pure gold, or was gold to which a small but significant quantity of copper had been added.

If we were to reconstruct Archimedes' reasoning, it might go as follows. Gold was the densest known substance (at that time). Its density in modern terms is 19·3 grams per cubic centimeter. This means that a given weight of gold takes up less volume than the same weight of anything else! In fact, a given weight of pure gold takes up less volume than the same weight of *any* kind of impure gold.

The density of copper is 8·92 grams per cubic centimeter, just about half that of gold. If we consider 100 grams of pure gold, for instance, it is easy to calculate it to have a volume of 5·18 cubic centimeters. But suppose the 100 grams of what looked like pure gold was really only 90 grams of gold and 10 grams of copper. The 90 grams of gold would have a volume of 4·66 cubic centimeters, while the 10 grams of copper would have a volume of 1·12 cubic centimeters; for a total value of 5·78 cubic centimeters.

The difference between 5·18 cubic centimeters and 5·78 cubic centimeters is quite a noticeable one, and would instantly tell if the crown were of pure gold, or if it contained 10 per cent copper (with the missing 10 per cent of gold tucked neatly in the goldsmith's strongbox).

All one had to do, then, was measure the volume of the crown and compare it with the volume of the same weight of pure gold.

The mathematics of the time made it easy to measure the volume of many simple shapes: a cube, a sphere, a cone, a cylinder, any flattened object of simple regular shape and known thickness, and so on.

We can image Archimedes saying, 'All that is necessary, sire, is to pound that crown flat, shape it into a square of uniform thickness, and then I can have the answer for you in a moment.'

Whereupon Hieron must certainly have snatched the crown away and said, 'No such thing, I can do that much without you; I've studied the principles of mathematics, too. This crown is a highly satisfactory work of art and I won't have it damaged. Just calculate its volume without in any way altering it.'

But Greek mathematics had no way of determining the volume of anything with a shape as irregular as the crown, since integral calculus had not yet been invented (and wouldn't be for two thousand years, almost). Archimedes would have had to say, 'There is no known way, sire, to carry though a non-destructive determination of volume.'

'Then think of one,' said Hieron testily.

And Archimedes must have set about thinking of one, and gotten nowhere. Nobody knows how long he thought, or how hard, or what hypotheses he considered and discarded, or any of the details.

What we do know is that, worn out with thinking, Archimedes decided to visit the public baths and relax. I think we are quite safe in saying that Archimedes had no intention of taking his problem to the baths with him. It would be ridiculous to imagine he would, for the public baths of a Greek metropolis weren't intended for that sort of thing.

The Greek baths were a place for relaxation. Half the social aristocracy of the town would be there and there was a great deal more to do than wash. One steamed one's self, got a massage, exercised, and engaged in general socializing. We can be sure that Archimedes intended to forget the stupid crown for a while.

One can envisage him engaging in light talk, discussing the latest news from Alexandria and Carthage, the latest scandals in town, the latest funny jokes at the expense of the country-squire Romans – and then he lowered himself into a nice hot bath which some bumbling attendant had filled too full.

The water in the bath slopped over as Archimedes got in. Did Archimedes notice that at once, or did he sigh, sink back, and paddle his feet awhile before noting the water-slop. I guess the latter. But, whether soon or late, he noticed, and that one fact, added to all the chains of reasoning his brain had been working on during the period of relaxation when it was unhampered by the comparative stupidities (even in Archimedes) of voluntary thought, gave Archimedes his answer in one blinding flash of insight.

Jumping out of the bath, he proceeded to run home at top speed through the streets of Syracuse. He did *not* bother to put on his clothes. The thought of Archimedes running naked through Syracuse has titillated dozens of generations of youngsters who have heard this story, but I must explain that the ancient Greeks were quite lighthearted in their attitude toward nudity. They thought no more of seeing a naked man on the streets of Syracuse, than we would on the Broadway stage.

And as he ran, Archimedes shouted over and over, 'I've got it! I've got it!' Of course, knowing no English, he was compelled to shout it in Greek, so it came out, '*Eureka! Eureka!*'

Archimedes' solution was so simple that anyone could understand it – once Archimedes explained it.

# THE EUREKA PHENOMENON

If an object that is not affected by water in any way, is immersed in water, it is bound to displace an amount of water equal to its own volume, since two objects cannot occupy the same space at the same time.

Suppose, then, you had a vessel large enough to hold the crown and suppose it had a small overflow spout set into the middle of its side. And suppose further that the vessel was filled with water exactly to the spout, so that if the water level were raised a bit higher, however slightly, some would overflow.

Next, suppose that you carefully lower the crown into the water. The water level would rise by an amount equal to the volume of the crown, and that volume of water would pour out the overflow and be caught in a small vessel. Next, a lump of gold, known to be pure and exactly equal in weight to the crown, is also immersed in the water and again the level rises and the overflow is caught in a second vessel.

If the crown were pure gold, the overflow would be exactly the same in each case, and the volumes of water caught in the two small vessels would be equal. If, however, the crown were of alloy, it would produce a larger overflow than the pure gold would and this would be easily noticeable.

What's more, the crown would in no way be harmed, defaced, or even as much as scratched. More important, Archimedes had discovered the 'principle of buoyancy'.

And was the crown pure gold? I've heard that it turned out to be alloy and that the goldsmith was executed, but I wouldn't swear to it.

How often does this 'Eureka phenomenon' happen? How often is there this flash of deep insight during a moment of relaxation, this triumphant cry of 'I've got it! I've got it!' which must surely be a moment of the purest ecstasy this sorry world can afford?

I wish there were some way we could tell. I suspect that in the history of science it happens *often*; I suspect that very few significant discoveries are made by the pure technique of voluntary thought; I suspect that voluntary thought may possibly prepare the ground (if even that), but that the final touch, the real inspiration, comes when thinking is under involuntary control.

But the world is in a conspiracy to hide that fact. Scientists are wedded to reason, to the meticulous working out of consequences from assumptions, to the careful organization of experiments designed to check those consequences. If a certain line of experiments ends nowhere, it is omitted from the final report. If an inspired guess turns out to be correct, it is *not* reported as an inspired guess. Instead, a solid line of voluntary thought is invented after the fact to lead up to the thought, and that is what is inserted in the final report.

The result is that anyone reading scientific papers would swear that *nothing* took place but voluntary thought maintaining a steady clumping stride from origin to destination, and that just can't be true.

It's such a shame. Not only does it deprive science of much of its glamour (how much of the dramatic story in Watson's *Double Helix* do you suppose got into the final reports announcing the great discovery of the structure of DNA?*), but it hands over the important process of 'insight', 'inspiration', 'revelation' to the mystic.

The scientist actually becomes ashamed of having what we might call a revelation, as though to have one is to betray reason – when actually what we call revelation in a man who has devoted his life to reasoned thought, is after all merely reasoned thought that is not under voluntary control.

Only once in a while in modern times do we ever get a glimpse into the workings of involuntary reasoning, and when we do, it is always fascinating. Consider, for instance, the case of Friedrich August Kekule von Stradonitz.

In Kekule's time, a century and a quarter ago, a subject of great interest to chemists was the structure of organic molecules (those associated with living tissue). Inorganic molecules were generally simple in the sense that they were made up of few atoms. Water molecules, for instance, are made up of two atoms of hydrogen and one of oxygen

* I'll tell you, in case you're curious. None!

($H_2O$). Molecules of ordinary salt are made up of one atom of sodium and one of chlorine (NaCl), and so on.

Organic molecules, on the other hand, often contained a large number of atoms. Ethyl alcohol molecules have two carbon atoms, six hydrogen atoms, and an oxygen atom ($C_2H_6O$); the molecule of ordinary cane sugar is $C_{12}H_{22}O_{11}$, and other molecules are even more complex.

Then, too, it is sufficient, in the case of inorganic molecules generally, merely to know the kinds and numbers of atoms in the molecule; in organic molecules, more is necessary. Thus, dimethyl ether has the formula $C_2H_6O$, just as ethyl alcohol does, and yet the two are quite different in properties. Apparently, the atoms are arranged differently within the molecules – but how to determine the arrangements?

In 1852, an English chemist, Edward Frankland, had noticed that the atoms of a particular element tended to combine with a fixed number of other atoms. This combining number was called 'valence'. Kekulé in 1858 reduced this notion to a system. The carbon atom, he decided (on the basis of plenty of chemical evidence) had a valence of four; the hydrogen atom, a valence of one; and the oxygen atom, a valence of two (and so on).

Why not represent the atoms as their symbols plus a number of attached dashes, that number being equal to the valence. Such atoms could then be put together as though they were so many Tinker Toy units and 'structural formulas' could be built up.

It was possible to reason out that the structural formula of ethyl alcohol was 
$$\begin{array}{cc} H & H \\ | & | \\ H-C-C-O-H \\ | & | \\ H & H \end{array}$$
, while that of dimethyl ether was 
$$\begin{array}{cc} H & H \\ | & | \\ H-C-O-C-H \\ | & | \\ H & H \end{array}$$

In each case, there were two carbon atoms, each with four dashes attached; six hydrogen atoms, each with one dash attached; and an oxygen atom with two dashes attached. The molecules were built up of the same components, but in different arrangements.

Kekule's theory worked beautifully. It has been immensely deepened and elaborated since his day, but you can still find structures very much like Kekule's Tinker Toy formulas in any modern chemical textbook. They represent oversimplifications of the true situation, but they remain extremely useful in practice even so.

The Kekule structures were applied to many organic molecules in the years after 1858 and the similarities and contrasts in the structures neatly matched similarities and contrasts in properties. The key to the rationalization of organic chemistry had, it seemed, been found.

Yet there was one disturbing fact. The well-known chemical benzene wouldn't fit. It was known to have a molecule made up of equal numbers of carbon and hydrogen atoms. Its molecular weight was known to be 78 and a single carbon-hydrogen combination had a weight of 13. Therefore, the benzene molecule had to contain six carbon-hydrogen combinations and its formula had to be $C_6H_6$.

But that meant trouble. By the Kekule formulas, the hydro-carbons (molecules made up of carbon and hydrogen atoms only) could easily be envisioned as chains of carbon atoms with hydrogen atoms attached. If all the valences of the carbon atoms were filled with hydrogen atoms, as in 'hexane', whose molecule looks like this—

```
    H H H H H H
    | | | | | |
H—C—C—C—C—C—C—H
    | | | | | |
    H H H H H H
```

the compound is said to be saturated. Such saturated hydrocarbons were found to have very little tendency to react with other substances.

If some of the valences were not filled, unused bonds were added to those connecting the carbon atoms. Double bonds were formed as in 'hexene'—

```
        H H H H H H
        | | | | | |
    H—C—C—C=C—C—C—H
        | |     | |
        H H     H H
```

Hexene is unsaturated, for that double bond has a tendency to open up and add other atoms. Hexene is chemically active.

When six carbons are present in a molecule, it takes fourteen hydrogen atoms to occupy all the valence bonds and make it inert – as in hexane. In hexene, on the other hand, there are only twelve hydrogens. If there were still fewer hydrogen atoms, there would be more than one double bond; there might even be triple bonds, and the compound would be still more active than hexene.

Yet benzene, which is $C_6H_6$ and has eight fewer hydrogen atoms than hexane, is *less* active than hexene, which has only two fewer hydrogen atoms than hexane. In fact, benzene is even less active than hexane itself. The six hydrogen atoms in the benzene molecule seem to satisfy the six carbon atoms to a greater extent than do the fourteen hydrogen atoms in hexane.

For heaven's sake, why?

This might seem unimportant. The Kekule formulas were so beautifully suitable in the case of so many compounds that one might simply dismiss benzene as an exception to the general rule.

Science, however, is not English grammar. You can't just categorize something as an exception. If the exception doesn't fit into the general system, then the general system must be wrong.

Or, take the more positive approach. An exception can

often be made to fit into a general system, provided the general system is broadened. Such broadening generally represents a great advance and for this reason, exceptions ought to be paid great attention.

For some seven years, Kekule faced the problem of benzene and tried to puzzle out how a chain of six carbon atoms could be completely satisfied with as few as six hydrogen atoms in benzene and yet be left unsatisfied with twelve hydrogen atoms in hexene.

Nothing came to him!

And then one day in 1865 (he tells the story himself) he was in Ghent, Belgium, and in order to get to some destination, he boarded a public bus. He was tired and, undoubtedly, the droning beat of the horses' hooves on the cobblestones, lulled him. He fell into a comatose half-sleep.

In that sleep, he seemed to see a vision of atoms attaching themselves to each other in chains that moved about. (Why not? It was the sort of thing that constantly occupied his waking thoughts.) But then one chain twisted in such a way that head and tail joined, forming a ring – and Kekule woke with a start.

To himself, he must surely have shouted 'Eureka', for indeed he had it. The six carbon atoms of benzene formed a ring and not a chain, so that the structural formula looked like this:

To be sure, there were still three double bonds, so you might think the molecule had to be very active – but now there was a difference. Atoms in a ring might be expected to have different properties from those in a chain and double bonds in one case might not have the properties of those in the other. At least, chemists could work on that assumption and see if it involved them in contradictions.

It didn't. The assumption worked excellently well. It turned out that organic molecules could be divided into two groups: aromatic and aliphatic. The former had the benzene ring (or certain other similar rings) as part of the structure and the latter did not. Allowing for different properties within each group, the Kekule structures worked very well.

For nearly seventy years, Kekule's vision held good in the hard field of actual chemical techniques, guiding the chemist through the jungle of reactions that led to the synthesis of more and more molecules. Then, in 1932, Linus Pauling applied quantum mechanics to chemical structure with sufficient subtlety to explain just why the benzene ring was so special and what had proven correct in practice proved correct in theory as well.

Other cases? Certainly.

In 1764, the Scottish engineer James Watt was working as an instrument maker for the University of Glasgow. The university gave him a model of a Newcomen steam engine, which didn't work well, and asked him to fix it. Watt fixed it without trouble, but even when it worked perfectly, it didn't work well. It was far too inefficient and consumed incredible quantities of fuel. Was there a way to improve that?

Thought didn't help; but a peaceful, relaxed walk on a Sunday afternoon did. Watt returned with the key notion in mind of using two separate chambers, one for steam only and one for cold water only, so that the same chamber did not have to be constantly cooled and reheated to the infinite waste of fuel.

The Irish mathematician William Rowan Hamilton worked up a theory of 'quaternions' in 1843 but couldn't complete that theory until he grasped the fact that there were conditions under which $p \times q$ was *not* equal to $q \times p$. The necessary thought came to him in a flash one time when he was walking to town with his wife.

The German physiologist Otto Loewi was working on the mechanism of nerve action, in particular, on the chemicals produced by nerve endings. He woke at 3 A.M. one night in 1921 with a perfectly clear notion of the type of experiment he would have to run to settle a key point that was puzzling him. He wrote it down and went back to sleep. When he woke in the morning, he found he couldn't remember what his inspiration had been. He remembered he had written it down, but he couldn't read his writing.

The next night, he woke againat 3 A.M. with the clear thought once more in mind. This time, he didn't fool around. He got up, dressed himself, went straight to the laboratory and began work. By 5 A.M. he had proved his point and the consequences of his findings became important enough in later years so that in 1936 he received a share in the Nobel prize in medicine and physiology.

How very often this sort of thing must happen, and what a shame that scientists are so devoted to their belief in conscious thought that they so consistently obscure the actual methods by which they obtain their results.

# 14 – POMPEY AND CIRCUMSTANCE

Rationalists have a hard time of it, because the popular view is that they are committed to 'explaining' everything.

This is not so. Rationalists maintain that the proper way of arriving at an explanation is through reason – but there is no guarantee that some particular phenomenon can be explained in that fashion at some given moment in history or from some given quantity of observation.*

Yet how often I (or any rationalist) am presented with something odd and am challenged, 'How do you explain that?' The implication is that if I don't explain it instantly to the satisfaction of the individual posing the question, then the entire structure of science may be considered to be demolished.

But things happen to me, too. One day in April 1967, my car broke down and had to be towed to a garage. In seventeen years of driving various cars, that was the first time I ever had to endure the humiliation of being towed.

When do you suppose the second time was? —Two hours later, on the same day, for a completely different reason.

Seventeen years without a tow, and then two tows on the same day! And how do you explain *that*, Dr. Asimov? (Gremlins? A vengeful Deity? An extraterrestrial conspiracy?)

On the second occasion, I did indeed loudly advance all three theories to my unruffled garageman. *His* theory (he was also a rationalist) was that my car was old enough to be falling apart. So I bought a new car.

---

* It is the mystics, really, who are committed to explaining *everything*, for they need nothing but imagination and words – *any* words, chosen at random.

Let's look at it this way! To every single person on Earth, a large number of events, great, small, and insignificant, happen each day. Every one of those events has some probability of occurrence, though we can't always decide the exact probability in each case. On the average, though, we might imagine that one out of every thousand events has an only one-in-a-thousand chance of happening; one out of every million events has an only one-in-a-million chance of happening; and so on.

This means that every one of us is constantly experiencing some pretty low-probability events. That is the normal result of chance. If any of us went an appreciable length of time with nothing unusual happening, that would be *very* unusual.

And suppose we don't restrict ourselves to one person, but consider, instead, all the lives that have ever been lived. The number of events then increases by a factor of some sixty billion and we can assume that sometime, to someone, something will happen that is sixty billion times as improbable as anything happening to some other particular man. Even such an event requires no explanation. It is part of our normal universe going along its business in a normal way.

Examples? We've all heard very odd coincidences that have happened to someone's second cousin, odd things that represent such an unusual concatenation of circumstance that surely we *must* admit the existence of telepathy or flying saucers or Satan or *something*.

Let me offer something, too. Not something that happened to my second cousin, but to a notable figure of the past whose life is quite well documented. Something very unusual happened to him, which in all my various and miscellaneous reading of history I have never seen specifically pointed out. I will, therefore, stress it to you as something more unusual and amazing than anything I have ever come across, and even so, it *still* doesn't shake my belief in the supremacy of the rational view of the universe. Here goes—

The man in question was Gnaeus Pompeius, who is better known to English-speaking individuals as Pompey.

## POMPEY AND CIRCUMSTANCE

Pompey was born in 106 B.C. and the first forty-two years of his life were characterized by uniform good fortune. Oh, I dare say he stubbed his toe now and then and got attacks of indigestion at inconvenient times and lost money on the gladiatorial contests – but in the major aspects of life, he remained always on the winning side.

Pompey was born at a time when Rome was torn by civil war and social turmoil. The Italian allies, who were not Roman citizens, rose in rebellion against a Roman aristocracy who wouldn't extend the franchise. The lower classes, who were feeling the pinch of a tightening economy, now that Rome had completed the looting of most of the Mediterranean area, were struggling against the senators, who had kept most of the loot.

When Pompey was in his teens, his father was trying to walk the tightrope. The elder Pompey had been a general who had served as consul in 89 B.C., and had defeated the Italian non-citizens and celebrated a triumph. But he was not an aristocrat by birth and he tried to make a deal with the radicals. This might have gotten him in real trouble, for he had worked himself into a spot where neither side trusted him, but in 87 B.C. he died in the course of an epidemic that swept his army.

That left young Pompey as a fatherless nineteen-year-old who had inherited enemies on both sides of the civil war.

He had to choose and he had to choose carefully. The radicals were in control of Rome, but off in Asia Minor, fighting a war against Rome's enemies, was the reactionary general Lucius Cornelius Sulla.

Pompey, uncertain as to which side would win, lay low and out of sight. When he heard that Sulla was returning victorious, from Asia Minor, he made his decision. He chose Sulla as probable victor. At once, he scrabbled together an army from among those soldiers who had fought for his father, loudly proclaimed himself on Sulla's side, and took the field against the radicals.

There was his first stroke of fortune. He had backed the right man. Sulla arrived in Italy in 83 B.C. and began winning at once. By 82 B.C. he had wiped out the last opposition in Italy and at once made himself dictator. For three years he was absolute ruler of Rome. He reorganized the government and placed the senatorial aristocrats firmly in control.

Pompey benefited, for Sulla was properly grateful to him. Sulla sent Pompey to Sicily, then to Africa, to wipe out the disorganized forces that still clung to the radical side there, and this was done without trouble.

The victories were cheap and Pompey's troops were so pleased that they acclaimed Pompey as 'the Great', so that he became Gnaeus Pompeius Magnus – the only Roman to bear this utterly un-Roman cognomen. Later accounts say that he received this name because of a striking physical resemblance between himself and Alexander the Great, but such a resemblance could have existed only in Pompey's own imagination.

Sulla ordered Pompey to disband his army after his African victories but Pompey refused to do so, preferring to stay surrounded by his loyal men. Ordinarily, one did not lightly cross Sulla, who had no compunctions whatever about ordering a few dozen executions before breakfast. Pompey, however, proceeded to marry Sulla's daughter. Apparently, this won Sulla over to the point of not only accepting the title of 'the Great' for the young man, but also to the point of allowing him to celebrate a triumph in 79 B.C. even though he was below the minimum age at which triumphs were permitted.

Almost immediately thereafter, Sulla resigned the dictatorship, feeling his work was done, but Pompey's career never as much as stumbled. He now had a considerable reputation (based on his easy victories). What's more, he was greedy for further easy victories.

For instance, after Sulla's death, a Roman general, Marcus Aemilius Lepidus, turned against Sulla's policies. The reactionary Senate at once sent an army against him. The senatorial army was led by Quintus Catulus, with Pompey as second-in-command. Until then, Pompey had supported Lepidus, but again he guessed the winning side in time.

## POMPEY AND CIRCUMSTANCE

Catulus easily defeated Lepidus, and Pompey managed to get most of the credit.

There was trouble in Spain at this time, for it was the last stronghold of radicalism. In Spain, a radical general, Quintus Sertorius, maintained himself. Under him, Spain was virtually independent of Rome and was blessed with an enlightened government, for Sertorius was an efficient and liberal administrator. He treated the native Spaniards well, set up a Senate into which they were admitted, and established schools where their young men were trained in Roman style.

Naturally, the Spaniards, who for some centuries had had a reputation as fierce and resolute warriors, fought heart and soul on the side of Sertorius. When Sulla sent Roman armies into Spain, they were defeated.

So, in 77 B.C., Pompey, all in a glow over Catulus's easy victory over Lepidus, offered to go to Spain to take care of Sertorius. The Senate was willing and off to Spain marched Pompey and his army. On his way through Gaul, he found the dispirited remnants of Lepidus' old army. Lepidus himself was dead by now but what was left of his men were under Marcus Brutus (whose son would, one day, be a famous assassin).

There was no trouble in handling the broken army and Pompey offered Brutus his life if he would surrender. Brutus surrendered and Pompey promptly had him executed. One more easy victory, topped by treachery, and Pompey's reputation increased.

On to Spain went Pompey. In Spain, a sturdy old Roman general, Metellus Pius, was unsuccessfully trying to cope with Sertorius. Vaingloriously, Pompey advanced on his own to take over the job – and Sertorious, who was the first good general Pompey had yet encountered, promptly gave the young man a first-class drubbing. Pompey's reputation might have withered then and there, but just in time, Metellus approached with reinforcements and Sertorious had to withdraw. At once, Pompey called it a victory, and, of course, got the credit for it. His luck held.

For five years, Pompey remained in Spain, trying to handle Sertorius, and for five years he failed. And then he had a stroke of luck, the luck that never failed Pompey, for Sertorius was assassinated. With Sertorius gone, the resistance movement in Spain collapsed. Pompey could at once win another of his easy victories and could then return to Rome in 71 B.C., claiming to have cleaned up the Spanish mess.

But couldn't Rome have seen it took him five years?

No, Rome couldn't, for all the time Pompey had been in Spain, Italy itself had been going through a terrible time and there had been no chance of keeping an eye on Spain.

A band of gladiators, under Spartacus, had revolted. Many dispossessed flocked to Spartacus' side, and for two years, Spartacus (a skillful fighter) destroyed every Roman army sent out against him and struck terror into the heart of every aristocrat. At the height of his power he had 90,000 men under his command and controlled almost all of southern Italy.

In 72 B.C., Spartacus fought his way northward to the Alps, intending to leave Italy and gain permanent freedom in the barbarian regions to the north. His men, however, misled by their initial victories, preferred to remain in Italy in reach of more loot. Spartacus turned south again.

The senators now placed an army under Marcus Licinius Crassus, Rome's richest and most crooked businessman. In two battles, Crassus managed to defeat the gladiatorial army and in the second one, Spartacus was killed. Then, just as Crassus had finished the hard work, Pompey returned with his Spanish army and hastily swept up the demoralized remnants. He immediately represented himself, successfully, as the man who had cleaned up the gladiatorial mess after having taken care of Spain. The result was that Pompey was allowed to celebrate a triumph, but poor Crassus wasn't.

The Senate, though, was growing nervous. They were not sure they trusted Pompey. He had won too many victories and was becoming entirely too popular.

Nor did they like Crassus (no one did). For all his wealth, Crassus was not a member of the aristocratic families and he grew angry at being snubbed by the socially superior Senate. Crassus began to court favor with the people with well-placed philanthropies. He also began to court Pompey.

Pompey always responded to courting and, besides, had an unfailing nose for the winning side. He and Crassus ran for the consulate in 70 B.C. (two consuls were elected each year), and they won. Once consul, Crassus began to undo Sulla's reforms of a decade earlier in order to weaken the hold of the senatorial aristocracy on the government. Pompey, who had been heart and soul with Sulla when that had been the polite thing to do, turned about and went along with Crassus, though not always happily.

But Rome was still in trouble. The West had been entirely pacified, but there was mischief at sea. Roman conquests had broken down the older stable governments in the East without having, as yet, established anything quite as stable in their place. The result was that piracy was rife throughout the eastern Mediterranean. It was a rare ship that could get through safely and, in particular, the grain supply to Rome itself had become so precarious that the price of food skyrocketed.

Roman attempts to clear out the pirates failed, partly because the generals sent to do the job were never given enough power. In 67 B.C., Pompey maneuvered to have himself appointed to the task – but under favorable conditions. The Senate, in a panic over the food supply, leaped at the bait.

Pompey was given dictatorial powers over the entire Mediterranean coast to a distance of fifty miles inland for three years and was told to use that time and the entire Roman fleet to destroy the pirates. So great was Roman confidence in Pompey that food prices fell as soon as news of his appointment was made public.

Pompey was lucky enough to have what no previous Roman had – adequate forces and adequate power. Nevertheless one must admit that he did well. In three *months*, not three years, he scoured the Mediterranean clear of piracy.

If he had been popular before, he was Rome's hero now.

The only place where Rome still faced trouble was in eastern Asia Minor, where the kingdom of Pontus had been fighting Rome with varying success for over twenty years. It had been against Pontus that Sulla had won victories in the East, yet Pontus kept fighting on. Now a Roman general, Lucius Licinius Lucullus, had almost finished the job, but he was a hard-driving martinet, hated by his soldiers.

When Lucullus' army began to mutiny in 66 B.C., just when one more drive would finish Pontus, he was recalled and good old Pompey was sent eastward to replace him. Pompey's reputation preceded him; Lucullus' men cheered him madly and for him did what they wouldn't do for Lucullus. They marched against Pontus and beat it. Pompey supplied the one last push and, as always, demanded and accepted credit for the whole thing.

All of Asia Minor was now either Roman outright or was under the control of Roman puppet governments. Pompey therefore decided to clean up the East altogether. He marched southward and around Antioch found the last remnant of the Seleucid Empire, established after the death of Alexander the Great two and a half centuries before. It was now ruled by a nonentity called Antiochus XIII. Pompey deposed him, and annexed the empire to Rome as the province of Syria.

Still further south was the kingdom of Judea. It had been independent for less than a century, under the rule of a line of kings of the Maccabean family. Two of the Maccabeans were now fighting over the throne and one appealed to Pompey.

Pompey at once marched into Judea and laid siege to Jerusalem. Ordinarily, Jerusalem was a hard nut to crack, for it was built on a rocky prominence with a reliable water supply; it had good walls; and it was usually defended with fanatic vigor.

Pompey, however, noticed that every seven days things were quiet. Someone explained to him that on the Sabbath,

the Jews wouldn't fight unless attacked and even then fought
without real conviction. It must have taken quite a while to
convince Pompey of such a ridicuous thing but, once convinced, he used a few Sabbaths to bring up his siege ma-
chinery without interference, and finally attacked on another
Sabbath. No problem.

Pompey ended the Maccabean kingdom and annexed
Judea to Rome while allowing the Jews to keep their
religious freedom, their Temple, their high-priests, and their
peculiar, but useful, Sabbath.

Pompey was forty-two years old at this time, and success
had smiled at him without interruption. I now skip a single
small event in Pompey's life and represent it by a line of
asterisks: one apparently unimportant circumstance.

\* \* \* \* \* \* \* \* \* \*

Pompey returned to Italy in 61 B.C. absolutely on top of
the world, boasting (with considerable exaggeration) that
what he had found as the eastern border of the realm he had
left at its center. He received the most magnificent triumph
Rome had ever seen up to that time.

The Senate was in terror lest Pompey make himself a dictator and turn to the radicals. This Pompey did not do.
Once, twenty years before, when he had an army, he kept
that army even at the risk of Sulla's displeasure. Now,
something impelled him to give up his army, disband it, and
assume a role as a private citizen. Perhaps he was convinced
that he had reached a point where the sheer magic of his
name would allow him to dominate the republic.

At last, though, his nose for the right action failed him.
And once having failed him, it failed him forever after.

To begin with, Pompey asked the Senate to approve everything he had done in the East, his victories, his treaties, his
depositions of kings, his establishment of provinces. He also
asked the Senate to distribute land to his soldiers, for he
himself had promised them land. He was sure that he had
but to ask and he would be given.

Not at all. Pompey was now a man without an army and

the Senate insisted on considering each individual act separately and nit-pickingly. As for land grants, that was rejected.

What's more, Pompey found that he had no one on his side within the government. All his vast popularity suddenly seemed to count for nothing as all parties turned against him for no discernible reason. What's more, Pompey could do nothing about it. Something had happened, and he was no longer the clever, golden-boy Pompey he had been before 64 B.C. Now he was uncertain, vacillating and weak.

Even Crassus was no longer his friend. Crassus had found someone else: a handsome, charming individual with a silver tongue and a genius for intrigue – a man named Julius Caesar. Caesar was a playboy aristocrat but Crassus paid off the young man's enormous debts and Caesar served him well in return.

While Pompey was struggling with the Senate, Caesar was off in Spain, winning some small victories against rebellious tribes and gathering enough ill-gotten wealth (as Roman generals usually did) to pay off Crassus and make himself independent. When he returned to Italy and found Pompey furious with the Senate, he arranged a kind of treaty of alliance between himself, Crassus, and Pompey – the 'First Triumvirate'.

But it was Caesar and not Pompey who profited from this. It was Caesar who used the alliance to get himself elected consul in 59 B.C. Once consul, Caesar controlled the Senate with almost contemptuous ease, driving the other consul, a reactionary, into house arrest.

One thing Caesar did was to force the aristocrats of the Senate to grant all of Pompey's demands. Pompey got the ratification of all of his acts and he got the land for his soldiers – and yet he did not profit from this. Indeed, he suffered humiliation, for it was quite clear that he was standing, hat in hand, while Caesar graciously bestowed largesse on him.

Yet Pompey could do nothing, for he had married Julia, Caesar's daughter. She was beautiful and winning and Pompey was crazy about her. While he had her, he could do nothing to cross Caesar.

Caesar was running everything now. In 58 B.C., he suggested that he, Pompey, and Crassus each have a province in which they could win military victories. Pompey was to have Spain: Crassus was to have Syria; and Caesar was to have southern Gaul, which was then in Roman hands. Each was to be in charge for five years.

Pompey was delighted. In Syria, Crassus would have to face the redoubtable Parthian kingdom, and in Gaul, Caesar would have to face the fierce-fighting barbarians of the North. With luck, both would end in disaster, since neither was a trained military man. As for Pompey, since Spain was quiet, he could stay in Italy and control the government. Who could ask for more?

It might almost seem that if Pompey reasoned this way, his old nose for victory had returned. By 53 B.C., Crassus' army was destroyed by the Parthians east of Syria and Crassus himself was killed.

But Caesar? No, Pompey's luck had *not* returned. To the astonishment of everyone in Rome, Caesar, who, until then, had seemed to be nothing but a playboy and intriguer, turned out, in middle age (he was forty-four when he went to Gaul), to be a first-class military genius. He spent five years fighting the Gauls, annexing the vast territory they inhabited, conducting successful forays into Germany and Britain. He wrote up his adventures in his *Commentaries* for the Roman reading public, and suddenly Rome had a new military hero. —And Pompey, sitting in Italy, doing nothing, was nearly dead of frustration and envy.

In 54 B.C., though, Julia died, and Pompey was no longer held back in his animus against Caesar. The senatorial aristocrats, now far more afraid of Caesar than of Pompey, flattered the latter, who promptly joined them and married a new wife, the daughter of one of the leading senators.

When Caesar returned from Gaul in 50 B.C., the Senate ordered him to disband his armies and enter Italy alone. It was clear that if Caesar did so, he would be arrested and probably executed. What, then, if he defied the Senate and brought his army with him?

'Fear not,' said Pompey, confidently, 'I have but to stamp my foot upon the ground and legions will rise up to support us.'

In 49 B.C., Caesar crossed the Rubicon River, which represented the boundary of Italy, and did so with his army. Pompey promptly stamped his foot – and nothing happened. Indeed, those soldiers stationed in Italy began to flock to Caesar's standards. Pompey and his senatorial allies were forced to flee, in humiliation, to Greece.

Grimly, Caesar and his army followed them.

In Greece, Pompey managed to collect a sizable army. Caesar, on the other hand, could only bring so many men across the sea and so Pompey now had the edge. He might have taken advantage of his superior numbers to cut Caesar off from his base and then stalk him carefully, without risking battle, and slowly wear him down and starve him out.

Against this was the fact that the humiliated Pompey, still dreaming of the old days, was dying to defeat Caesar in open battle and show him the worth of a *real* general. Worse yet, the senatorial party insisted on a battle. So Pompey let himself be talked into one; after all, he outnumbered Caesar two to one.

The battle was fought at Pharsalus in Thessaly on June 29, 48 B.C.

Pompey was counting on his cavalry in particular, a cavalry consisting of gallant young Roman aristocrats. Sure enough, at the start of the battle, Pompey's cavalry charged round the flank of Caesar's army and might well have wreaked havoc from the rear and cost Caesar the battle. Caesar, however, had foreseen this and had placed some picked men to meet the cavalry, with instructions not to throw their lances but to use them to poke directly at the faces of the horsemen. He felt that the aristocrats would not stand up to the danger of being disfigured and he was right. The cavalry broke.

With Pompey's cavalry out, Caesar's hardened infantry

broke through the more numerous but much softer Pompeian line and Pompey, unused to handling armies in trouble, fled. In one blow, his entire military reputation was destroyed and it was quite clear that it was Caesar, not Pompey, who was the real general.

Pompey fled to the one Mediterranean land that was not yet entirely under Roman control – Egypt. But Egypt was in the midst of a civil war at the time. The boy-king, thirteen-year-old Ptolemy XII, was fighting against his older sister, Cleopatra, and the approach of Pompey created a problem. The politicians supporting young Ptolemy dared not turn Pompey away and earn the undying enmity of a Roman general who might yet win out. On the other hand, they dared not give him refuge and risk having Caesar support Cleopatra in revenge.

So they let Pompey land – and assassinated him.

And that was the end of Pompey, at the age of fifty-six.

Up to the age of forty-two he had been uniformly successful; nothing he tried to do failed. After the age of forty-two he had been uniformly unsuccessful; nothing he tried to do succeeded.

What happened at the age of forty-two? What circumstance took place in the interval represented earlier in the article by the line of asterisks that might 'explain' this. Well, let's go back and fill in that line of asterisks.

* * * * * * * * * *

We are back in 64 B.C.

Pompey is in Jerusalem, curious about the queer religion of the Jews. What odd things do they do besides celebrate a Sabbath? He began collecting information.

There was the Temple, for instance. It was rather small and unimpressive by Roman standards but was venerated without limit by the Jews and differed from all other temples in the world by having no statue of a god or goddess inside. It seemed the Jews worshipped an invisible god.

'Really?' said the amused Pompey.

Actually, he was told, there was an innermost chamber

in the Temple, the Holy of Holies, behind a veil. No one could ever go beyond the veil but the high priest, and he could only do so on the Day of Atonement. Some people said that the Jews secretly worshiped an ass's head there, but of course, the Jews themselves maintained that only the invisible presence of God was in that chamber.

Pompey, unimpressed by superstition, decided there was only one way of finding out. He would look inside this secret chamber.

The high priest was shocked, the Jews broke into agonized cries of dismay, but Pompey was adamant. He was curious and he had his army all around him. Who could stop him? So he entered the Holy of Holies.

The Jews were undoubtedly certain that he would be struck by lightning or otherwise destroyed by an offended God, but he wasn't.

He came out again in perfect health. He had found nothing, apparently, and nothing had happened to him, *apparently*.*

*In case you think I'm turning mystical myself, please reread the introduction to this chapter.

## 15 – BILL AND I

I am, as it happens, doing a book on Lord Byron's narrative poem *Don Juan*.* The poem is an uninhibited satire in which Byron takes the opportunity to lash out at everything and everyone that displeased him. He is cruel to the point of sadism toward Britain's monarchs, toward its poet laureate, toward its greatest general, and so on.

But those for whom he reserves his most savage sallies are his critics. Byron did not take to criticism kindly and he invariably struck back.

Now, as far as I know, there is no such thing as a writer who takes no criticism kindly. Most of us, however, affect stoic unconcern and bleed in private.

For myself, alas, stoic unconcern is impossible. My frank and ingenuous countenance is a blank page on which my every emotion is clearly written (I am told) and I don't think I have ever succeeded in playing the stoic for even half a a second. Indeed, when I am criticized unfairly, everyone within earshot knows that I have been – and for as much as hours at a time.

Naturally, when I recently published a two-volume book entitled *Asimov's Guide to Shakespeare* I tried to steel myself for inevitable events. It was bound to get into the hand of an occasional Shakespearian scholar who would come all over faint at the thought of someone outside their field daring to invade the sacred precincts.

In fact, the very first review I received began: 'What is Isaac Asimov, spinner of outer-space tales, doing—'

Naturally, I read no further. The fact that I am a spinner of outer-space tales is utterly irrelevant to this particular

* Because I want to, and because my publishers humor me.

book and can only be mentioned because the reviewer thinks there is something vaguely (or not so vaguely) beneath literary dignity in being a science fiction writer.

I have sought a printable response for that and failed, so I'll pass on.

A second review was much more interesting. It appeared in a Kentucky paper and was written by someone I will call Mr. X. It begins this way: 'Isaac Asimov is associate professor of biochemistry at Boston University, and a prolific writer in many fields. I have read several of his books on science with the greatest attention and respect.'

So far, so good. I am delighted.

But then, a very little while later, he says: 'In this book, however, he has left the sunlit paths of natural science for the treacherous bogs of literature—'

What he objects to, it seems, is that I have annotated the plays, explained all the historical, legendary, and mythological references. It is a book of footnotes, so to speak, and he resents it. He points out that he thinks of 'the language, the poetry, as the chief glory of Shakespeare's works'.

Well, who doesn't? I'm delighted that Mr. X is clever enough to understand the language and the poetry without any explanation from me. And if *he* doesn't need it, why should anyone else, eh?

Notice, though, that he doesn't scorn to follow me along 'the sunlit paths of natural science'. Indeed, he reads my books on science 'with the greatest attention and respect'.

I'm glad he does and I can only presume that he is grateful that I take the trouble to footnote science so that he can get a fugitive hint of its beauties.

Suppose, instead, I were to say to Mr. X, 'The logarithm of two is a transcendental number; and, indeed, the logarithm of any integer to any integral base is transcendental except where the integer is equal to the base or to a power of that base.'

Mr. X might then, with justification, say, 'What is a transcendental number, a logarithm, and, in this case, a power and a base?'

In fact, if he were a really deep thinker, he might ask, 'What is two?'

But suppose I answered then that my statement bore within it all the poetry and symmetry and beauty of mathematics ('Euclid alone has looked on beauty bare') and that to try to explain it would simply hack it up. And if Mr. X found trouble in understanding it, too bad for him. He just wasn't as bright as I was, and he could go to blazes.

But I *don't* answer that way. I explain such matters and many more, and go to a lot of trouble to do so, and then he reads those explanations with 'the greatest attention and respect'.

Scientists generally recognize the importance of explaining science to the non-scientist. It is interesting, then, in a rather sad way, that there exist humanists who feel themselves to be proprietors of their field, who hug literature to themselves, who mumble 'the language, the poetry', and who see no reason why it should be explained to anyone as long as they themselves can continue to sniff the ambrosia.

Let us take a specific case. In the last act of *The Merchant of Venice*, Lorenzo and Jessica are enjoying an idyllic interlude at Portia's estate in Belmont, and Lorenzo says:

> *Sit, Jessica. Look how the floor of heaven*
> *Is thick inlaid with patens of bright gold.*
> *There's not the smallest orb which thou behold'st*
> *But in his motion like an angel sings,*
> *Still quiring to the young-eyed cherubins;*
> *Such harmony is in immortal souls,*
> *But whilst this muddy vesture of decay*
> *Doth grossly close it in, we cannot hear it.*

I think this passage is beautiful, for I have as keen a sense of the beauty and poetry of words as Mr. X; perhaps (is it possible?) even keener.

But what happens if someone says: 'What are patens?'

After all, it is not a very common word. Does it ruin the

beauty of the passage to whisper in an aside, 'Small disks'?

Or what if someone asks, 'What does he mean about these orbs singing like angels in their motion? What orbs? What motion? What singing?'

Am I to understand that the proper answer is, 'No! Just listen to the words and the beautiful flow of language, and don't ask such philistine questions.'

Does it spoil the beauty of Shakespeare's language to understand what he is saying? Or can it be that there are humanists who, qualified though they may be in esthetics, know little of the history of science, and don't know what Bill is saying and would rather not be asked?

All right, then, let's use this as a test case. I am going to explain this passage in far greater detail than I did in my book, just to show how much there is to consider in these beautiful syllables—

Anyone looking at the sky in a completely unsophisticated manner, without benefit of any astronomical training whatever, and willing to judge by appearances alone, is very likely to conclude that the Earth is covered by a smooth and flattened dome of some strong and solid material that is blue by day and black by night.

Under that solid dome is the air and the floating clouds. Above it, he may decide, is another world of gods and angels where the immortal souls of men will rise after the body dies and decays.

As a matter of fact, this is precisely the view of the early men of the Near East, for instance. On the second day of creation, says the Bible: 'God said, Let there be a firmament in the midst of the waters, and let it divide the waters from the waters. And God made the firmament, and divided the waters which were under the firmament from the waters which were above the firmament' (Genesis 1: 6–7).

The word 'firmament' is from the Latin word *firmamentum*, which means something solid and strong. This is a translation of the Greek word *stereoma*, which means something solid and strong, and that is a translation of the original Hebrew word *raqia*, which refers to a thin metallic bowl.

In the biblical view there was water below the firmament (obviously) and water above it, too, to account for the rain. That is why in the time of Noah's flood, it is recorded that '... the fountains of the great deep [were] broken up, and the windows of heaven were opened' (Genesis 7:11). The expression might be accepted as metaphor, of course, but I'm sure that the unsophisticated accepted it literally.

But there's no use laughing from the height of our own painfully gained hindsight. About 700 B.C., when the material of Genesis was first being collected, the thought that the sky was a solid vault with another world above it was a reasonable conclusion to come to from the evidence available.

What's more, it would seem reasonable about 700 B.C. to suppose that the firmament stretched over but a limited portion of a flat Earth. One could see it come down and join the Earth tightly at the horizon. Few people in ancient times ever traveled far from home and the world to them was but a few miles in every direction. Even soldiers and merchants, who tramped longer distances, might feel the Earth was larger than it looked but that the world to the enlarged horizon was still flat, and still enclosed on all sides by the junction of firmament and ground. (This was also very much the medieval view and probably that of many unsophisticated moderns.)

The Greek philosophers, however, had come to the conclusion, for a number of valid reasons, that the Earth was *not* a more or less flat object of rather limited size, but a spherical object of sufficient size to dwarf the known world to small dimensions.

The firmament, then, must stretch all around the globular Earth, and to do so symmetrically, it must be another, but much larger, sphere. The apparent flattening of the firmament overhead had to be an illusion (it is!) and the Greeks spoke of what we would call the 'heavenly sphere' as opposed to the 'terrestrial sphere'.

None of this, however, altered the concept of the firmament (or heavenly sphere) as made up of something hard and firm. What, then, were the stars?

Naturally, the first thought was that the stars were exactly what they appeared to be: tiny, glowing disks embedded in the material of the firmament ('Look how the floor of heaven is thick inlaid with patens of bright gold').

The evidence in favor of this was that the stars did not fall down, as they would surely do if they were not firmly fixed to the heavenly sphere. Secondly, the stars moved about the Earth once every twenty-four hours, with the North Star as one pivot (the other being invisible behind the southern horizon), and did so all in one piece without altering their relative positions from night to night and from year to year.

If the stars were suspended freely somewhere between the heavenly sphere and the Earth, and for some reason did not fall, surely they would either not move at all or, if they did, would move independently. No, it made much more sense to suppose them all fixed to the heavenly sphere, and to suppose that it was the heavenly sphere that turned, carrying all the stars with itself.

But alas, this interpretation of the heavens – beautiful and austerely simple – did not account for everything.

As it happened, the Moon was clearly not imbedded in the heavenly sphere, for it did not maintain a fixed position relative to the stars. It was at a particular distance from a particular star one night, farther east the next night, still farther east the one afterward. It moved steadily west to east in such a way as to make a complete circuit of the starry sky in a little over twenty-seven days.

The Sun moved from west to east, too, relative to the stars, though much more slowly. Its motion couldn't be watched directly, of course, since no stars were visible in its neighborhood by which its position might be fixed. However, the nighttime configuration of stars shifted from night to night because, clearly, the Sun moved and blotted out slightly different portions of the sky from day to day. In that manner it could be determined that the Sun seemed to make a circuit of the sky in a little over 365 days.

If the Sun and the Moon were the only bodies to be exceptional, this might not be too bad. After all, they were very much different from the stars and could not be expected to follow the same rules.

Thus the Hebrews, in their creation myth, treated the Sun and Moon as special cases. On the fourth day of creation, 'God made two great lights; the greater light to rule the day, and the lesser light to rule the night: he made the stars also' (Genesis 1: 16).

It seems amusing to us today to have the stars dismissed in so offhanded a fashion, but it makes perfect sense in the light of the Hebrew knowledge of the day. The stars were all imbedded in the firmament and they served only as a background against which the motions of the Sun and the Moon could be studied.

But then it turned out that certain of the brighter stars were also anomalous in their motions and shifted positions against the background of the other stars. In fact, their motion was even stranger than that of the Moon and the Sun, for, though they moved west to east most of the time, relative to the stars, as the Moon and the Sun did, they occasionally would turn about and move east to west. Very puzzling!

The Greeks called these stars *planetes*, meaning 'wanderers', as compared with the 'fixed stars'. The Greek word has become 'planet' to us and seven of them were recognized. These included the five bright stars which we now call Mercury, Venus, Mars, Jupiter, and Saturn, and, of course, the Sun and the Moon.

What to do with them? Well, like the stars, the planets did not fall and like the stars they moved about the Earth. Therefore, like the stars, they had to be embedded in a sphere. Since each of the seven planets moved at a different speed and in a different fashion, each had to have a separate sphere, one nested inside the other, and all nested inside the sphere of the stars.

Thus there arose the notion not of the heavenly sphere, but of the heavenly spheres, plural.

But there was only one heavenly sphere that could be seen – the blue sphere of the firmament. The fact that the other spheres were invisible was no argument, however, for their non-existence, merely for their transparency. They were sometimes called 'the crystalline spheres', where the word 'crystalline' was used in its older meaning as 'transparent'.

The Greeks then set about trying to calculate where the different spheres were pivoted and how they must turn in order to cause each planet to move in the precise fashion in which it was observed to move. Endless complications had to be added in order to match theory with observation, but for two thousand years the complicated theory of the crystalline spheres held good, not because men of thought were perversely stupid, but because nothing else so well fit the appearances.

Even when Copernicus suggested that the Sun, not the Earth, was the center of the universe, he didn't abolish the spheres. He merely had them surrounding the Sun, with the Earth itself embedded in one of them. It was only with Johannes Kepler—

But never mind that. The details of the motions of the crystalline spheres don't concern us in this article. Let us instead consider an apparently simpler question: In what order are the spheres nested? If we were to travel outward from Earth, which sphere would we come to first, which next, and so on.

The Greeks made the logical deduction that the closest sphere would be smallest and would therefore make a complete turn in the briefest time. Since the Moon made a complete circle against the stars in about four weeks (a far shorter time than any other planet managed to run the course), its sphere must be closest.

Arguing in this manner, the Greeks decided the next closest sphere was that of Mercury; then, in order, Venus, the Sun, Mars, Jupiter, and Saturn. And finally, of course, there was the sphere of the stars.

And how far apart were the spheres and what were their actual distances from the Earth?

That, unfortunately, was beyond the Greeks. To be sure, the Greek astonomer Hipparchus, about 150 B.C., used a perfectly valid method (after the still earlier astronomer Aristarchus) for determining the distance of the Moon, and had placed it at a distance of thirty times the Earth's diameter, which is correct, but the distance of no other heavenly body was determined with reasonable accuracy until the seventeenth century.

Now the scene switches. About 520 B.C., the Greek philosopher Pythagoras was plucking strings, and found that he could evoke notes that harmonized well together if he used strings whose lengths were simply related. One string might be twice the length of another; or three strings might have lengths that were in the ratio of 3:4:5.

The details are irrelevant, but to Pythagoras it seemed highly significant that there should be a connection between pleasing sounds and small whole numbers. It fit in with his rather mystical notion that everything in the universe was related to simple ratios and numbers.

Those who followed in his footsteps after his death accentuated the mysticism and it seemed to the Pythagoreans that they now had a way of deciding not only the how of planets, but the why as well. Since numbers governed the universe, one ought to be able to deduce the way in which the universe ought to be constructed.

For instance, 10 was a particularly impressive number. (Why? Well, for one thing, $1+2+3+4 = 10$, and this seems to have some mystical value.) In order, then, for the universe to function well, it had to be composed of ten spheres.

Of course, there were only eight spheres, one for the stars and one for each of the seven planets, but that didn't stop the Pythagoreans. They decided that the Earth moved around some central fire of which the Sun was only a reflection, and worked up a reason for explaining why the central fire was invisible. That added a ninth sphere for the Earth. In addition, they imagined another planet on the opposite side of the central fire, a 'counter-Earth'. The counter-Earth kept pace with the Earth and stayed always beyond the central fire and was thus never seen. Its sphere was the tenth.

In addition, the Pythagoreans thought that the spheres were nested inside each other in such a way that their distances of separation bore simple ratios to one another and produced harmonious notes in their motion as a result (like the plucking of strings of simply related lengths). Originally, I imagine, the Pythagoreans may have advanced this notion of harmonious notes only as a metaphor to represent the simply related distances, but later mystics accepted the notes as literally existent. They became 'the music of the spheres'.

Of course, no one ever heard any music from the sky, so it had to be assumed to be inaudible to men on Earth. It is this notion that causes Shakespeare to speak of an orb that 'in his motion like an angel sings' but with sounds that can be heard only in heaven ('Such harmony is in immortal souls'). While men's souls are still draped in their earthly bodies, they are deaf to it ('whilst this muddy vesture of decay doth grossly close it in, we cannot hear it').

Well, then, does understanding Lorenzo's speech in terms of ancient astronomy spoil its beauty? Does it not seem that to understand him adds to the interest? Does it not remove the nagging question of 'But what does it mean?' that otherwise distracts from an appreciation of the passage?

It may be, of course, that Mr. X is the kind who never asks 'But what does it mean?' It may be that for him understanding is irrelevant. If so, he and I are not soul mates. It may even be that Mr. X is the kind of obscurantist who finds that understanding decreases beauty. If so, he and I are even less soul mates.

And yet, let me point out that there is something in this very passage that could be of interest to Shakespearian scholars *if* they thoroughly understood what Shakespeare was talking about.

As almost everyone knows, there are many who feel that Shakespeare did not write the plays attributed to him. They feel that someone else did, with the person most frequently credited being Francis Bacon, who was an almost exact contemporary of Shakespeare.

The argument very often heard is that Shakespeare was just a fellow from the provinces with very little education and that he could not possibly have written so profoundly learned a set of plays. Bacon, on the other hand, was a great philosopher and one of the most intensely educated people of his time. Bacon, therefore, could easily have written the plays.

Shakespearian scholars, when they argue the matter at all, are forced to maintain that Shakespeare was much better educated than he is given credit for being and that therefore he was learned enough to write his plays. Since virtually nothing is known of Shakespeare's life, the argument will never be settled in that fashion.

Why not turn matters around, then, and argue that Bacon was *too* educated to write Shakespeare's plays, that there exist errors in the plays that Bacon could never possibly have made and that would just suit an insufficiently educated fellow from the sticks?

Consider Lorenzo's speech. Lorenzo is talking about the *stars*; these are the 'patens of bright gold' with which 'the floor of heaven is thick inlaid.' Lorenzo (hence Shakespeare) seems to think that each star has a separate sphere and that each gives out its own note ('There's not the smallest orb which thou behold'st/But in his motion like an angel sings').

Lest you think I'm misinterpreting the speech, let's take a clearer case.

In Act II of *A Midsummer Night's Dream*, Oberon is reminding Puck of a time they listened to a mermaid who sang with such supernal beauty that

> ... the rude sea grew civil at her song,
> And certain stars shot madly from their spheres,
> To hear the sea maid's music.

The use of the plural 'spheres' shows again that Shakespeare thinks that each star has its separate sphere.

This is wrong. There is a sphere for each planet; one for the Earth itself, if you like; one for the counter-Earth; one for any imaginary planet you wish. However, all the ancient theories agreed that the 'fixed stars' were all embedded in a single sphere.

To imagine separate spheres for each star, as Shakespeare does more than once in his plays, is to display a lack of knowledge of Greek astronomy. This is a lack of knowledge that Francis Bacon could not possibly have displayed; hence we might fairly argue that Francis Bacon could not possibly have written Shakespeare's plays.

Well, don't get me wrong. I don't want to imply that I received only bad reviews for my *Guide to Shakespeare*. Actually, most of the reviews were quite complimentary and were an entire pleasure to read.

Just the same, I had better start preparing myself for the occasional review by the 'outraged specialist' type that I will surely get when *Asimov's Annotated 'Don Juan'* is published.

*F – The Problem of Population*

## 16 – STOP!

As some of my Gentle Readers may know, I am an after-dinner speaker when I can be persuaded to be one. (For the information of prospective persuaders, I may as well state at once that the best persuasion is a large check.)

As a speaker, I must be introduced, of course, and introductions vary in quality. It's not difficult to see that a short introduction is better than a long one, since much preliminary talk dulls the edge of the audience and makes the speaker's task harder.

Again, a dull introduction is better than a witty one, since a speaker can easily suffer by contrast with preliminary wit, and an audience which might otherwise be receptive enough becomes critical after the joy of the introduction.

Needless to say, then, the very worst possible introduction a speaker can have is one that is both long and witty, and on the night of April 20, 1970, at Pennsylvania State University, that is exactly what I got.

Phil Klass (far better known to science fiction fans as William Tenn) is associate professor of English at Penn State and it naturally fell to him to introduce me. With an evil smile on his face, he got up and delivered an impassioned address that went on for fifteen minutes and that had the audience of some twelve hundred people rocking with laughter (at my expense, naturally). As he went on, a kind of grimness settled about my soul. I couldn't possibly follow him; he was too good. Naturally, I decided to kill him as soon as I got my hands on him, but first I had to live through my own talk.

And then at the very last minute, Phil (I'm sure, unintentionally) saved me. He concluded his talk by saying,

'But don't let me give you the idea that Asimov is a Renaissance Man. He has never, after all, sung *Rigoletto* at the Metropolitan Opera.'

I brightened up at once, rose smiling from my seat, and mounted the stage. I waited for the polite opening applause to die down and, without preliminary, launched my resonant voice into '*Bella figlia dell'amore—*' the opening of the famous Quartet from *Rigoletto*.

It was the first time I ever got the biggest laugh of the entire evening with my first four words, and after that I had no trouble at all.

I tell you all this because in April 1970, I gave nine talks which, despite *Rigoletto*, were not funny at all. It was the month in which the first Earth Day was celebrated, and every one of my talks dealt, in whole or in part, with the coming catastrophe.

I have discussed that catastrophe in the final chapters of a previous volume, *The Stars in Their Courses* (Doubleday, 1971), and I have made it quite plain that in my opinion the first order of business is a halt to the population increase on Earth. Without such a halt right away, none of mankind's problems can be solved under any conditions: *none*!

The question then is: How can the population increase be halted?

Since this is now the prime question and, indeed, the only relevant question that futurists have to face, and since science fiction writers were futurists long before the word was invented, and since I am self-admittedly one of the leading science fiction writers, I consider it my duty to try to answer this question.

To begin with, let us admit there are only two general ways of bringing about a halt in the population increase: we might increase the death rate, or we might decrease the birth rate. (We might, conceivably, do both, but the two are independent and can be discussed separately.)

Let's start with the increase of the death rate first and consider all the variations on the theme:

## A – Increase in the death rate

### 1 – Natural increase

This is the system that has been in use for all species since life began. It is the system that served to limit human population throughout its history. When food grew scarce, human beings starved to death, were easier prey for disease in their famished conditions, fought each other and killed in order to gain access to what food supplies there were, led armies into other regions where food was more plentiful. For all these reasons the death rate rose precipitously and population fell to match the food supply.

We have here the 'four horsemen of the Apocalypse' (see the sixth chapter of the biblical Book of Revelation) – war, civil strife, famine, and pestilence.

Modern science has greatly weakened the force of the third and fourth horsemen, and both famine and pestilence are not what they once were. This in itself has amazingly lowered the death rate from what it was in all the millennia before 1850 and is the major reason for the explosiveness with which population has increased since.

We can well imagine, however, that if the population continues to soar for another generation, the efforts of science will crack under the strain. All four horsemen will regain their ascendancy; the death rate will zoom upward.

Possibly one might be objective about this and say: Well, this is the way the game of life is played. The fittest will survive and mankind will continue stronger than ever, for the winnowing-out it has received.

Not at all! There might have been some validity to this view, for all its inhumanity, if mankind were armed with stone axes and spears, or even with machine guns and tanks. Unfortunately, we have nuclear weapons at our disposal and when the four horsemen start out on their horrid ride, the H-bombs will surely be used.

Mankind, living in the tattered remnants of a world torn by thermonuclear war, will *not* be stronger than ever. It will be living not only in the ruins of a destroyed technology, but in the midst of a dangerously poisoned soil, sea, and atmosphere which may no longer be able to support vertebrate life at all.

We'll need something better.

## 2 – Directed general increase

### a – Involuntary

Instead of waiting for the course of events to enforce a catastrophic increase in death rate, we might blow off steam by randomly killing off part of the population from year to year. Suppose that preliminary estimates during a census year make it seem that the world population is 10 per cent above optimum. In that case, take the census and shoot every tenth person counted.*

About the only thing that can be said about this method is that it is perhaps a little better than a thermonuclear war. I don't think any sane man would consider it if any other alternative existed at all.

### b – Voluntary

Random killing might be made voluntary if one constructed a suicide-centered society.† In such a society, suicide must be made to seem attractive, either through the effective promise of an afterlife or through the more material offer of financial benefits to the family left behind.

Somehow, though, I doubt that under any persuasion not involving physical constraint or emotional inhumanity, enough people will kill themselves to halt the population increase. Even if enough did, the kind of society that would place the accent on death with sufficient firmness to bring it about would undoubtedly be too unbearably morbid for the health of the species.

---

\* 'The Census Takers', an excellent science fiction story by Frederik Pohl, actually uses this situation.

† Gore Vidal's *Messiah* had something of this sort.

## 3 – Directed special increase

### a – Inferiority

But if we must kill, would it be possible to neutralize some of the horror by making murder serve some useful purpose. Suppose we kill off or (more humanely) sterilize that portion of the population that contributes least to mankind, the 'inferior' portion, in other words.

Indeed, such a policy has been put into practice on numerous occasions, though not usually out of a set, reasoned-out population strategy. Throughout Earth's history, a conquering nation has usually made the calm assumption that its own people were superior to the conquered people, who were therefore killed or enslaved as a matter of course. Under conditions of famine the conquered peasantry would surely die in greater proportion than the conquering aristocracy.

Conquerors varied in inhumanity. In ancient times, the Assyrians were most noted for the callous manner in which they would destroy the entire male population of captured cities; and in medieval times, the Mongols made a name for themselves in the same fashion. In modern times, the Germans under Hitler, more consciously and deliberately, set about destroying those whom they considered members of inferior races.

This policy can never be popular except with those who have the power and the inhumanity to declare themselves superior (and not usually with all of those either). The majority of mankind is bound to be among the conquered and the inferior and their approval is not to be expected. The Assyrians, Mongols, and Nazis were all greeted with nearly universal execration both in their own times and thereafter.

There are individuals whom the world generally would consider inferior – the congenital idiot, the psychopathic murderer, and so on – but the numbers of such people are too few to matter.

## b – *Old age*

Perhaps then people can be killed off according to some category that isn't as subjective as superiority-inferiority. What about the very old? They still eat; they are still drains on the culture; yet they give back very little.

There have been cultures which killed those aged members that could not carry their own weight (the Eskimos, for instance). Before late modern times, however, there was usually little pressure in this direction, since very few members of a society managed to live long enough to be too old to be worth their keep. Indeed, the very few aged members might even be valuable as the repositories of tradition and custom.

Not so nowadays. With the rise in life expectancy to seventy, the 'senior citizen' is far more numerous in absolute numbers and in proportion than ever before. Ought all those who reach sixty-five, say, to be painlessly killed? If this applies to all humans without exception there would be no subjective choice and no question of superiority-inferiority.

But what good would it do? The men and women thus killed are past the child-bearing age and have already done their damage. Such euthanasia will make the population younger but not do one thing to stop the population increase.

## c – *Infants*

Then why not the other end of the age scale? Why not kill babies? Infanticide has been a common enough method of population control in primitive societies, and in some not so primitive. Usually, it is the girl babies that are allowed to die, and, to be sure that is as it should be.

I hasten to say that I do not make the last statement out of anti-female animus. It is just that it is the female who is the bottleneck. Compare the female, producing thirteen eggs a year and fertile for limited periods each month, with the male, producing millions of sperm each day and nearly continuously on tap. A hundred thousand women will produce the same number of babies a year whether there are ten thousand men at their free disposal or a million men.

Actually, there are some points in favor of infanticide. For one thing, it definitely works. Carried out with inhuman efficiency, it could put an end to the human race altogether in the space of a century. It can be argued moreover that a newborn baby is only minimally conscious and doesn't suffer the agonies of apprehension; that he as yet lacks personality and that no emotional ties have had a chance to form about him.

And yet, infanticide isn't pleasant. Babies are helpless and appealing and a society that can bring itself to slaughter them is perhaps too callous and inhumane to serve mankind generally. Besides, we cannot kill all babies, only some of them, and at once an element of choice enters. Which babies? The Spartans killed all those that didn't meet their standards of physical fitness and in general the matter of superiority-inferiority enters with all its difficulties.

### d – *Fetuses*

What about pre-birth infanticide – in short, abortion. Fetuses are not independently living and society's conscience might be quieted by maintaining they are therefore not truly alive. They are not killed, they are merely 'aborted', prevented from gaining full life.

Of all forms of raising the death rate, abortion would seem the least inhumane, the least abhorrent. At the present moment, in fact, there are movements all over the world, and not least in the United States to legalize abortion.

And yet if one argues that killing a baby is not quite as bad as killing a grown man, and killing a fetus not quite as bad as killing a baby, why not go one step farther, and kill the fetus at the very earliest moment? Why not kill it before it has become a fetus, before conception has taken place?

It seems to me then that any humane person, considering all the various methods of raising the death rate must end by deciding that the best method is to prevent conception; that is, to lower the birth rate. Let's consider that next.

If we consider the different ways of decreasing the birth

rate, we can see that, to begin with, they fall in two broad groups: voluntary and involuntary.

## B – Decrease in birth rate

### 1 – Voluntary

Ideally, this is the situation most acceptable to a humane person. If the population increase must be halted, let everyone agree to and voluntarily practice the limitation of children.

Everyone might simply agree to have no more than two children. It would be one, then two, then STOP!

If this came to pass, not only would the population increase come to a halt* it would begin to decrease. After all, not all couples would have two children. Some, through choice or circumstance, would have only one child and some even none at all. Furthermore, of the babies that were born, some would be bound to die before having a chance to become adults and have babies of their own.

With each generation under the two-baby system, then, the total population of mankind would decrease substantially.

I do not consider this a bad thing at all, for I feel that the Earth is already, at this moment, seriously overpopulated. I could argue and have, that a closer approach to the ideal population of Earth would be one billion people, and this goal would allow several generations of shrinkage. In a rational society, without war or threat of war, it seems to me that a billion people could be supported indefinitely.

If the population threatened to drop below a billion, it would be the easiest thing in the world to raise the permitted number of babies to three per couple. Enough couples would undoubtedly take advantage of permission to have a third child to raise the population quickly.

---

* Provided the life expectancy doesn't increase drastically. If it did, there would be a continued accumulation of old people. It might be just as well not to labor to increase that expectancy above the level that now exists. It embarrasses me to say so but I see no way out.

I would anticipate that under a humane world government, a decennial census applied to the whole world would, on each occasion, serve to guide the decision whether, for the next ten years, third children would be asked for or not.

Such a system would work marvelously well, if it were adopted, but would it be? Would individuals limit births voluntarily? I am cynical enough to think not.

In the first place, where two is the desired number of babies per couple, it is so much easier to far overshoot the mark than far undershoot it. A particular couple can, without biological difficulty, have a dozen children, ten above par. No couple, however, no matter how conscientious can have fewer than zero children, or two under par.

This means that for every socially unfeeling couple with a dozen children, five couples must deprive themselves of children altogether to redress the balance.

Furthermore, I suspect that those families who, on a strictly voluntary basis, choose to have many children, are apt to be drawn from those with less social consciousness, less feeling of responsibility – for whatever reason. Each generation will contribute to the next generation in a most unbalanced fashion.

This would, in fact, very likely cause an utter breakdown in the voluntary system in short order, for there will be resentment and fear on the part of the socially conscious. The socially conscious will easily convince themselves that it is precisely the ignorant, the inferior, the undeserving who are breeding and they may feel that it is important for them to supply the world with their own, much-more-desirable offspring.

It is even rather likely that, as long as birth control is purely voluntary, it will be negated out of local subplanetary considerations.

In Canada, for instance, the birth rate is higher among the French-speaking portion of the population than among the English-speaking portion. I am sure that there are those on both sides of the fence who calculate, with hope or with fear, that the French-Canadians will eventually dominate the land out of sheer natural increase.

The French-Canadians might be loath to adopt voluntary birth control and lose the chance of domination, while English-Canadians might be loath to adopt it and perhaps hand over the domination all the more quickly to a still breeding French-Canadian population.

The situation might be similar within the United States, where Blacks have a higher birth rate than whites; or in Israel, where the Arabs have a higher birth rate than the Jews; or in almost any country with a non-homogeneous population.

It is not only inside a country where such questions would arise. The Greeks would not want to fall too far behind the Bulgarians in population; the Belgians too far behind the Dutch; the Indians too far behind the Chinese; and so on and so on.

Each nation, each group within a nation, would watch its neighbors and would attempt to retain the upper hand for itself or (which is the same thing) prevent the neighbor from gaining the upper hand. And, in the name of patriotism, nationalism, racism, voluntary birth control would fail and mankind would be doomed.

## 2 – Involuntary

Ought we then not merely to ask couples not to have more than two children; ought we to *tell* them?

Suppose, for instance, that all babies were carefully registered and that every time a woman had a second baby, the first one being still alive, she be routinely sterilized before being released from the hospital.

Why women? you might ask. Why not men, for whom the operation is simpler.

My choice of women is not the result of male chauvinism on my part but only because women, as I said before, are the bottleneck in reproduction. Sterilizing some males will do no good if the rest merely work harder at it, while sterilizing females *must* force the birth rate down. Then, too, one knows when a female has two children; one can only guess at it with males. Finally, it is the woman, not the man, who is on the hospital table at the time of birth.

But would such involuntary birth control work? Or would it arouse such resentment that the world would constantly rock with insurrection, that women would have their babies in secret, that the government would be forced into more extremes of tyranny constantly.

Somehow I suspect that the system would indeed break down if the process were not carried through without exception.

There would be a strong temptation, I suppose, to work out some sort of regulations whereby some people would be allowed three children or even four, while others might be allowed only one or even none at all. You might argue that college graduates ought to have more children than morons should; proven achievers, more than idle dreamers; athletes, more than diabetics: and so on.

Unfortunately, I don't think that any graduated system, however impartially and sensibly carried through, can possibly succeed.

Whatever the arrangement, there will be an outcry that group X is favored over group Y. At least group Y will say so and will gather information to prove that group X is in control of the World Population Council. Using the same statistics and information, group X will insist that group Y is being favored.

The only possible solution, however wasteful, would be to allow no exceptions at all for any reason. Let the 'fit' have no more children than the 'unfit' (no less, either), in whatever way your own emotions and prejudices happen to define 'fit' and 'unfit'.

Then, when the population is reduced to the proper level and the Earth has had several generations of experience with a humane world government, propositions for grading birth numbers and improving the quality of humanity without increasing its quantity may be entertained.

Yet I must admit that the use of the knife, the inexorable push of governmental surgery is unpalatable to me and would probably be unpalatable to many people. If there were only some way to make voluntary compliance as surefire as the involuntariness of sterilization, I would prefer that.

Could we leave people the choice; could we let them choose the additional child if they wish – but make it prohibitive for various reasons? Could we find pressures as inexorable as the knife, yet leaving the human body and, therefore, human dignity intact?

*3 – Voluntary, with encouragement*

Let's go back to voluntary birth limitation, but now let's not make it entirely voluntary. Let's set up some stiff penalties for lack of co-operation.

To begin with, reverse the philosophy of the income tax. At present births are encouraged by income tax deductions. Suppose there are penalties instead. Your tax would go up slightly with one child, up again slightly with two, and then up prohibitively with three.

In other words, couples are bribed not to have children.

There are other forms of bribes. When a third child is born, a husband might suffer a pay cut, or lose his job altogether and be forced to go on welfare. A three-child family may lose medical plan privileges, be barred from air flight, be ostracized by other families.

This is all very cruel but in the world today that third child is a social felony.

Is that kind of pressure better than the knife? Will it force mankind less strongly into secret births, whole hidden colonies of forbidden children? Will the third children who *are* born be mistreated or killed? Will the rule discriminate in favor of the rich?

I don't know, but I can't think of anything better. It seems to me that the need is overwhelming and the time is now. Let's begin at once to persuade people, one way or another, not to have babies, to begin building the social pressures against large families. It is that, or the death of civilization and of billions of human beings with it.*

Only one thing—

Suppose we adopt this final alternative and suppose humanity generally and genuinely accepts it. People everywhere honestly intend to have no more than two children. Each couple which has its two children must now decide (without compulsory sterilization, mind you) to figure out a way not to have the third.

How? What alternatives are open to them? —For remember, if there are no reasonable alternatives, we are back to compulsory sterilization. —Or doom.

* In case your curiosity has grown unbearable, I myself have two children. I will have no more.

# 17 – ... BUT HOW?

Sometimes I wish I were smart enough to know when I've happened to say something smart so that I can get it down on paper and notarize it, as proof for posterity.

For instance, back in 1952, I was listening to the news of the election-day Eisenhower landslide with considerable gloom* when a ray of sunshine penetrated the darkness.

It seemed a young Democrat had just won his election to the Senate by a comfortable margin in the face of the tidal wave in the other direction at the presidential level. He was shown thanking his election workers and, in doing so, displayed such irresistible charm that I turned to my wife and said:

'If he weren't a Catholic, he'd be the next President of the United States, after Eisenhower.'

You're ahead of me, I know, but that young man was John F. Kennedy and I was remarkable prescient. Unfortunately, I have no record of the remark and my wife – the only witness – doesn't remember it.

On the other hand, at about the same time, in the early 1950's, I said, in the course of a discussion at a social gathering, 'This is the last generation in which the unrestricted right to breed will remain unquestioned. After this, birth control will be enforced.'

'What about the Roman Catholic Church?' someone asked me.

'The Roman Catholic Church,' I said, 'will have no choice but to go along.'

I was hooted down by unanimous consensus and it was the general feeling that being a science fiction writer had

* I will hide nothing from you. I am a Democrat.

gone to my head – but I still stand on what I said nearly twenty years ago.

So we'll limit births for reasons I explained in the previous chapter.

—But how?

There are many methods of birth control practiced. There is abstention and chastity, for example. (Don't laugh! For some people, this works, and we are in no position to turn down the help offered by any method, however minimal.) There is the rhythm method, of choosing, or trying to choose, that time of the month when a woman is not ovulating. There is the practice of withdrawal, or of surgical and permanent sterilization, or of chemical and temporary sterilization, or of mechanical interception, and so on.

All have their value as far as birth control is concerned; all have their disadvantages; no one method will do the trick by voluntary acceptance; perhaps even all together will not do the trick.

Nevertheless, we must try, and if anyone can think of some technique that is not being tried but ought to be, it is his duty, in this crisis facing mankind to advance it as forcefully as he can. This I intend to do.

The real enemy, as I see it, is social pressure, which is the strongest human force in the world. Love laughs at locksmiths and may flourish under the severest legal condemnations, but it is love indeed that can persist under no punishment worse than the cold-hearted ostracism of society.

Social pressure is irrepressible. The rebels who stand firmly against the Establishment and who object to all the moss-grown mores of yore, quickly develop a subculture with mores of its own which they do not, and dare not, violate.

And it is social pressure, inexorable social pressure, that dictates that people shall have children – lots of children – the more children the better.

There is reason for it. Despite what many think, the conventions of society are not invented merely to annoy and confuse, or out of a perverse delight in stupidity. They make sense – in the context of the times in which they originate.

Until the nineteenth century, there was virtually no place on Earth and virtually no time in history in which life expectancy was greater than thirty-five years. In most places and most times it was considerably less. There was virtually no place and no time in which infant mortality wasn't terrifyingly high. It was not the death of children that was surprising, but their survival.

Through all the ages of high infant mortality and low life expectancy, it stood to reason that each family had to have as many children as possible. This was not because each family sat down and worried about the future of mankind in the abstract. Not at all; it was because in a tribal society, the family is the social and cultural unit, and as many young as possible were necessary to carry on the work of herding or farming or whatever, while standing to their weapons to keep off other tribes at odd moments. And it took all the children the women could have to supply the necessary manpower.

With death so prevalent through hunger, disease, and warfare, the problem of overpopulation did not arise. If, unexpectedly, a tribe's numbers did increase substantially, they could always move outward and fall on the next tribe. It was the withering and extinction of the tribe that seemed the greater danger.

Consequently, social pressures were in favor of children, and naturally and rightly so.

We needn't go off into anthropological byways to see evidence of this; we have it at our fingertips in the Bible – the most important single source of social pressure in Western civilization. (And this is crucial, for it is Western culture that controls the Earth militarily, and Western culture that will have to lead the way in population policy.)

The first recorded statement of God to humanity after its creation is: 'And God blessed them and said unto them, Be fruitful and multiply, and replenish the Earth—' (Genesis 1: 28).

On a number of occasions thereafter, the Bible records

the fact that the inability to bear children is considered an enormous calamity. God promises Abram that he will be taken care of, saying, '. . . Fear not, Abram: I am thy shield and thy exceeding great reward' (Genesis 15: 1). But Abram can find no comfort in this and says, '. . . Lord God, what wilt thou give me, seeing I go childless . . .' (Genesis 15: 2).

In fact, childlessness was viewed as divine punishment. Thus, Jacob married two sisters: Leah and Rachel. He had wanted only Rachel but had been forced to take Leah through a trick. As a result, he showed considerable favoritism and of this God apparently disapproved; 'And when the Lord saw that Leah was hated, he opened her womb: but Rachel was barren' (Genesis 29: 31).

Naturally, Rachel was upset. 'And when Rachel saw that she bare Jacob no children, Rachel envied her sister; and said unto Jacob, Give me children, or else I die' (Genesis 30: 1).

There is the case of Hannah, who was barren, despite constant prayer; and who was miserable over it, despite the faithful love of her husband, who overlooked her barrenness (which made her worthless in a tribal sense and which placed her under strong suspicion of sinfulness) and expressed his love for her most touchingly: 'Then said Elkanah her husband to her, Hannah, why weepest thou? and why eatest thou not? and why is thy heart grieved? am not I better to thee than ten sons?' (1 Samuel 1: 8).

But Hannah perseveres in prayer and conceives at last, bearing Samuel. The second chapter of the book contains her triumphant song of celebration.

A particularly clear indication that barrenness is the punishment of sin arises in connection with the history of David. David had brought the Ark of the Covenant into Jerusalem and, in celebration, had participated in the ritualistic, orgiastic dance of celebration, one in which (the Bible is not clear) there may have been strong fertility-rite components. David's wife, Michal, disapproved strongly, saying sarcastically, '. . . How glorious was the king of Israel today, who uncovered himself today in the eyes of the handmaids of his servants, as one of the vain fellows shamelessly uncovereth himself!' (2 Samuel 6: 20).

This criticism displeased David and, apparently, God as well, for 'Therefore Michal the daughter of Saul had no child unto the day of her death' (2 Samuel 6: 23).

So strong was the tribal push for children that if a wife were barren, she herself might take the initiative of forcing her husband to impregnate a servant of her own, that she might have the credit of children by surrogate. Thus, when Abram's wife, Sarai, proved barren, she said to her husband, '... Behold now, the Lord hath restrained me from bearing: I pray thee, go in unto my maid; it may be that I may obtain children by her ...' (Genesis 16: 2).

Similarly, Jacob's wife, Rachel, lent her husband her maid, Bilhah, while his other wife, Leah, not to be behindhand, made her maid Zilpah available. These four women, among them, are described as being the mothers of the various ancestors of the twelve tribes of Israel.

It worked the other way, too. If a husband died before having children, it was the duty of the nearest member of the family (the brother, if possible) to make the effort of impregnating the widow in order that she might have sons which would then be counted to the credit of the dead man.

Thus, Jacob's fourth son, Judah, had an oldest son, Er, for whom he arranged a marriage with a young lady named Tamar. Unfortunately, Er died, so Judah told his next son, Onan: '... Go in unto thy brother's wife and marry her, and raise up seed to thy brother' (Genesis 38: 8).

Onan, however, did not want to. 'And Onan knew that the seed should not be his; and it came to pass, when he went in unto his brother's wife, that he spilled it on the ground, lest that he should give seed to his brother. / And the thing which he did displeased the Lord: wherefore he slew him also' (Genesis 38: 9–10).

Thus, the sin of Onan is not masturbation (which is what the word 'onanism' means) but what we call 'coitus interruptus.'

The pressure to bear children exists because a tribal society would not long survive without converting women into baby machines, and the biblical tales reflect this.

To be sure, there are religious sects which glorify birth control – in the form of chastity and virginity – but almost invariably because they expect the imminent end of the Earth.* The early Christians were among these and to this day chastity is a Christian virtue, and virginity is considered a pretty praiseworthy thing. Yet, even so, it is taken for granted in our traditional society that the greatest fulfillment a woman can possibly experience on Earth is that of becoming a wife and mother, that motherhood is of all things on Earth the most sacred, that to have many children is really a blessing and to have few children, or none, through some act of will, is somehow to be selfish.

The pressures produce important myths about men, too, for to have many children seems to be accepted as proving something about a man's virility. Even today, the father of triplets or more sometimes manages a look of smug modesty before the camera, an 'oh-it-was-nothing' expression that he thinks befits the sexual athlete. (Actually, whatever a man does or does not do has no connection at all with multiple births.)

All these pressures inherited from the dead past exist, then, despite the fact that the situation is now no longer what it was in tribal days. It is completely and catastrophically the opposite. We no longer have an empty Earth, we have a full one. We no longer have a short life expectancy, but a long one. We no longer have a high infant mortality rate, but a low one. We are no longer doubling Earth's population in several millennia, but in several decades.

Yet when we speak of birth control even today, we still have to overcome all the age-old beliefs of the tribal situation. Clearly, social pressure can be fought only with social pressure and as an example I have sometimes suggested

* Which, in a way, is why the modern population experts are pushing for birth control, too, because otherwise they expect the imminent end of the Earth.

(with a grin, lest I be lynched on the spot) that we begin by abolishing Mother's Day and replacing it with Childless Day, in which we honor all the adult women without children.

Social pressures involves more than merely a question of having children or not having children. The social pressures that for thousands of years have insisted on children, have gone into detail to make sure that these children come to pass. They have definitely and specifically outlawed the easiest methods of birth control, methods which require no equipment, no chemicals, no calculations, no particular self-control, methods which, if applied, *under tribal conditions of yore*, would have threatened the tribe with extinction.

So successful has this pressure been that such methods of birth control have passed beyond human ken, apparently. At least, when I hear proponents of birth control speak, or read what they write, I never seem to hear or see any mention of these natural methods. Either they are blissfully ignorant of them, or are afraid to speak of them.

The fact is, you see, that there are a variety of sexual practices that seem to give satisfaction, that do no physiological harm, and that offer no chance, whatsoever, for conception.

One and all, these stand condemned in our society for reasons that stretch back to the primitive necessity for babies.

For instance, the simplest possible non-conception-centered sexual practice is masturbation (in either male or female). It reduces tension and does no physiological harm.

Yet for how many years in our own society has it been viewed as an unspeakable vice (despite the fact that I understand, it is almost universally practiced). The pressure to consider it as more than a vice, and as actually a sin, has been such that in the effort to find biblical thunder against it, Onan's deed was considered masturbation, which it most certainly was not.

Clearly, the real crime of masturbation is that it wastes semen which, by tribal views, ought to be used in a sporting effort to effect conception. To say this, however, would be alien to the spirit of our society, so lies are invented instead. Masturbation (the threat goes) 'weakens' you; by which is meant that you won't perform effectively with women – a horrifying possibility to most men. Worse than that is the wild threat that masturbation gives rise to degeneracy (whatever that is) and even insanity.

Actually, it does none of these things. It does not even have the evils implicit in its being a 'solitary vice.' It can be indulged in, in company, and not necessarily in 'vile orgies,' but in ordinary heterosexual interaction.

All the strictures and fulminations against masturbation have never succeeded in wiping it out. It continued universal. What the lies did do, however, was to force the act to be carried on in secret, in shame, and in fear, so that those lies helped raise up generations of neurotics with distorted and utterly unnecessary hangups about sex. And why? To pay lip service to practices necessary to primitive tribes, but *fatal* to ourselves.

Part and parcel of the battle against masturbation is that against pornography. There have been periods in history when pornography was driven underground with scorn and disgust. This did not wipe out 'dirty books', 'dirty pictures', and 'dirty jokes'. It lent them an added titillation, if anything, But the drive against pornography did make it clear that sex was filthy, and therefore utterly distorted the attitude of millions concerning an activity which is both necessary and intensely pleasurable.

And what is the reason usually given for forbidding pornography? The one I hear most often is that it will inflame minds and cause people encountering such 'filth' to go ravening out into the street like wild beasts, seeking to rape and pervert.

It is ridiculous to think so. I suspect that what happens when you involve yourself with pornography, assuming it succeeds in arousing 'vile impulses' within you, is that you masturbate at the first opportunity. It releases tension rather than building it.

It is, in fact, by building tensions through a studied effort to consider sex dirty and forbidden, that one is most likely to be driven to rape.

No, the real evil of pornography in a tribal society is that, by encouraging masturbation, it diminishes the chance of conception.

There is a whole array of practices which, by the society and therefore by law, are stigmatized as perverse, as unnatural, as unspeakable, as 'crimes against nature', and so on. That these are unnatural is clearly not so, for if they were they would be easy to suppress. Indeed, there would be no need to suppress them, for they wouldn't exist. It is unnatural, for instance, to fly by flapping your arms, so that there are no laws against it. It is unnatural to live without breathing, so no one has to inveigh against it.

What is true about the so-called perversions is that they are *very* natural. They are so natural, indeed, that not all the shackles of the law, and not all the hellfire of religion, can serve to wipe them out.

And what harm do they do? Are they sicknesses?

I frequently hear homosexuality spoken of as a sickness, for instance, and yet there have been societies in which it was taken more or less for granted. Homosexuality was prevalent, and even approved, in the Golden Age of Athens; it was prevalent during the Golden Age of Islam; and despite everything, it was prevalent (I understand) among the upper classes of the Victorian Age.

It may be sickness but it does not seem to be inconsistent with culture. And how much of its sickness is the result of the hidden world in which it is forced to live, the fear and shame that are made to accompany it?

What is the real crime of all these so-called perversions? Might it not be that one and all are effective birth control agents. No practicing, exclusive homosexual, male or female, can possibly make or become pregnant. No one can ever impregnate or be impregnated by oral-genital contacts.

So what's wrong – in a time when birth rate *must* be lowered?

I don't mean that there aren't practices that *do* do harm, and these one ought to oppose. Sadomasochistic practices carried beyond the level of mild stimulation are not to be encouraged, for the same reason we oppose mutilation and murder. Those practices which involve seriously unhygienic conditions should be discouraged for the same reason any other unhygienic condition is discouraged.

Nor do I imply that we must *force* people to practice perversions.

I, for instance, am not a homosexual and wouldn't consider becoming one just to avoid having children. Nor would I persuade anyone to become one for that purpose and that purpose only.

I merely say that in a world threatened by overconception, it is useless and even suicidally harmful to carry on a battle against those who, of their own accord, prefer homosexuality, who in doing so do us no harm, and who, indeed, spare us children. Furthermore, there are borderline cases who might be homosexuals if left to themselves; shall we force them, by unbearable social pressure, into loveless heterosexual marriages, and into presenting the world with unneeded babies?

How do we justify this in the endangered world of the late twentieth century?

Social pressure – and the law – invades the bedrooms of even legally married individuals and dictates their private sexual practices. I am told that if a man and wife wish to practice anal or oral intercourse and are caught at it, they can be given stiff jail sentences in almost any state of the Union.

Why? What harm have they done themselves or anyone else? It is punishment without crime.

The 'harm', or course, is that they've practiced a completely effective birth control method that requires no equipment, no preparation, and supplies them, presumably, with satisfaction – something incompatible with the needs of a long-dead-and-gone tribal past.

I have heard it said that the practice of 'perversions' is

'corrupting', that it replaces the 'normal way', which is then neglected.

I've never seen evidence presented to back this view, but even if it were true, what then? What is the 'normal way' in a world like ours which must dread overconception? And if someone doesn't like the 'normal way' and therefore doesn't have children, whose business is that? If that same couple chose not to have children by practicing abstention, would anyone care? Would the law care? Then what's wrong with not having children another way? Because pleasure is a crime?

In David Reuben's book *Everything You Always Wanted to Know about Sex*, he devotes a section to oral-genital contacts, of which he seems to approve, but concludes that 'regular copulation is even more enjoyable'.

Actually, I suppose that is for each individual to decide for himself, but even if 'regular copulation' *is* more enjoyable, what then? If you find roast beef more enjoyable than bread and butter, is that a reason to outlaw bread and butter? And if you can't have roast beef and must choose between bread and butter and starvation, would you choose starvation?

It might very well be that it is variety that is best of all, and that for law and custom to try to insist on a monotony which, of all monotonies, is most dangerous to us today, is the greatest perversion of all.

Let's summarize, then.

I think that the importance of birth control is such that we ought to allow no useful method to lie unused.

All the common methods have their drawbacks: abstention is nearly impossible; sterilization is abhorrent; the rhythm method is cold-blooded and deprives the female of sex at just the time of the month she is most receptive; mechanical devices slow you up just when you least want to slow up; chemicals are bound to have side effects. I think, then, there is room for another method, particularly one which has none of these drawbacks.

With that in mind I think that social pressure against those practices commonly called 'perversions' ought to be lifted, where these are not physiologically harmful. The very qualities that made them perversions in a conception-centered society make them virtues in a non-conception-centered society.

I think that sex education ought to include not only information concerning what is usually considered 'normal' but also about those practices which are non-conception-centered. No one need to be taught to indulge in them exclusively, but by knowing they exist and aren't 'wrong', the number of occasions that so-called normal intercourse need be indulged in, with all the complications and drawbacks of artificial birth control methods, can be reduced. And, of course, if a couple have no children, and want one or two, they will know what to do.

As for those who can't stomach 'perversions' and who insist on doing everything by the numbers in the way that was good enough for their grandmother (I wonder!), then good luck to them, but they had better be careful.

One way or another, birth control must be made effective, and what I have suggested here is only one more method; one which, joined to the others already available, increases by that much the general effectiveness of the system as a whole and makes the chance just a little bit greater that the world might yet be saved.

**Isaac Asimov, Grand Master of Science Fiction, in Panther Books**

*The Foundation Trilogy*
| | |
|---|---|
| FOUNDATION | 40p ☐ |
| FOUNDATION AND EMPIRE | 40p ☐ |
| SECOND FOUNDATION | 40p ☐ |
| | |
| THE EARLY ASIMOV (Volume 1) | 40p ☐ |
| THE EARLY ASIMOV (Volume 2) | 35p ☐ |
| THE EARLY ASIMOV (Volume 3) | 40p ☐ |
| | |
| THE GODS THEMSELVES | 50p ☐ |
| THE CURRENTS OF SPACE | 40p ☐ |
| THE NAKED SUN | 40p ☐ |
| THE STARS LIKE DUST | 35p ☐ |
| THE CAVES OF STEEL | 40p ☐ |
| THE END OF ETERNITY | 35p ☐ |
| EARTH IS ROOM ENOUGH | 40p ☐ |
| THE MARTIAN WAY | 50p ☐ |
| NIGHTFALL ONE | 30p ☐ |
| NIGHTFALL TWO | 30p ☐ |
| ASIMOV'S MYSTERIES | 35p ☐ |
| | |
| I, ROBOT | 40p ☐ |
| THE REST OF THE ROBOTS | 50p ☐ |

*Edited by Asimov*
| | |
|---|---|
| NEBULA AWARD STORIES 8 | 60p ☐ |

**Panther Science Fiction – A Selection from the World's Best S.F. List**

| | | |
|---|---|---|
| GREYBEARD | Brian W. Aldiss | 40p ☐ |
| THE MOMENT OF ECLIPSE | Brian W. Aldiss | 35p ☐ |
| THE DISASTER AREA | J. G. Ballard | 30p ☐ |
| DO ANDROIDS DREAM OF ELECTRIC SHEEP? | Philip K. Dick | 30p ☐ |
| NOW WAIT FOR LAST YEAR | Philip K. Dick | 50p ☐ |
| CLANS OF THE ALPHANE MOON | Philip K. Dick | 50p ☐ |
| THE ZAP GUN | Philip K. Dick | 40p ☐ |
| ALL THE SOUNDS OF FEAR | Harlan Ellison | 30p ☐ |
| THE TIME OF THE EYE | Harlan Ellison | 35p ☐ |
| THE RING OF RITORNEL | Charles L. Harness | 35p ☐ |
| THE CENTAURI DEVICE | M. John Harrison | 50p ☐ |
| THE MACHINE IN SHAFT TEN | M. John Harrison | 50p ☐ |
| THE VIEW FROM THE STARS | Walter M. Miller, Jr. | 35p ☐ |
| MASQUE OF A SAVAGE MANDARIN | Philip Bedford Robinson | 35p ☐ |
| THE MULLER-FOKKER EFFECT | John Sladek | 35p ☐ |
| THE STEAM-DRIVEN BOY | John Sladek | 35p ☐ |
| LET THE FIRE FALL | Kate Wilhelm | 35p ☐ |
| BUG-EYED MONSTERS | Edited by Anthony Cheetham | 40p ☐ |

**More Great Science Fiction Books from Panther**

|  |  |  | |
|---|---|---|---|
| DOUBLE STAR | Robert A. Heinlein | 40p | ☐ |
| BEYOND THIS HORIZON | Robert A. Heinlein | 40p | ☐ |
| STRANGE RELATIONS | Philip José Farmer | 35p | ☐ |
| THE LEFT HAND OF DARKNESS | Ursula K. LeGuin | 50p | ☐ |
| CITY OF ILLUSIONS | Ursula K. LeGuin | 35p | ☐ |
| THE LATHE OF HEAVEN | Ursula K. LeGuin | 35p | ☐ |
| THE FREDERIK POHL OMNIBUS | Frederik Pohl | 40p | ☐ |
| THE WONDER EFFECT | Frederik Pohl & C. M. Kornbluth | 40p | ☐ |

*The Classic LENSMAN series*

|  |  |  | |
|---|---|---|---|
| TRIPLANETARY | E. E. 'Doc' Smith | 50p | ☐ |
| FIRST LENSMAN | E. E. 'Doc' Smith | 40p | ☐ |
| GALACTIC PATROL | E. E. 'Doc' Smith | 50p | ☐ |
| GREY LENSMAN | E. E. 'Doc' Smith | 50p | ☐ |
| SECOND STAGE LENSMEN | E. E. 'Doc' Smith | 50p | ☐ |
| CHILDREN OF THE LENS | E. E. 'Doc' Smith | 40p | ☐ |
| MASTERS OF THE VORTEX | E. E. 'Doc' Smith | 35p | ☐ |

*All these books are available at your local bookshop or newsagent, or can be ordered direct from the publisher. Just tick the titles you want and fill in the form below.*

Name..........................................................

Address.......................................................

..............................................................

Write to Panther Cash Sales, PO Box 11, Falmouth, Cornwall TR10 9EN

Please enclose remittance to the value of the cover price plus:
UK: 18p for the first book plus 8p per copy for each additional book ordered to a maximum charge of 66p

BFPO and EIRE: 18p for the first book plus 8p per copy for the next 6 books, thereafter 3p per book

OVERSEAS: 20p for the first book and 10p for each additional book

Granada Publishing reserve the right to show new retail prices on covers, which may differ from those previously advertised in the text or elsewhere.

LAH

**SPECIAL MESSAGE TO READERS**

This book is published under the auspices of
**THE ULVERSCROFT FOUNDATION**
(registered charity No. 264873 UK)
Established in 1972 to provide funds for
research, diagnosis and treatment of eye diseases.
Examples of contributions made are: —

A Children's Assessment Unit at
Moorfield's Hospital, London.

•

Twin operating theatres at the
Western Ophthalmic Hospital, London.

•

A Chair of Ophthalmology at the
Royal Australian College of Ophthalmologists.

•

The Ulverscroft Children's Eye Unit at the
Great Ormond Street Hospital For Sick Children,
London.

You can help further the work of the Foundation
by making a donation or leaving a legacy. Every
contribution, no matter how small, is received
with gratitude. Please write for details to:

**THE ULVERSCROFT FOUNDATION,
The Green, Bradgate Road, Anstey,
Leicester LE7 7FU, England.
Telephone: (0116) 236 4325**

**In Australia write to:
THE ULVERSCROFT FOUNDATION,
c/o The Royal Australian and New Zealand
College of Ophthalmologists,
94-98 Chalmers Street, Surry Hills,
N.S.W. 2010, Australia**